Vision Facts

Vision Facts
Questions about the Human Eye

Jason Yang & Dr. Charles Pidgeon

Sehej Bindra, Arnav Rastogi,
Tirtho Banerjee, Yash Gupta

Universal Publishers
Irvine • Boca Raton

Vision Facts: Questions about the Human Eye

Copyright © 2018 Charles Pidgeon

Universal Publishers, Inc.
Irvine • Boca Raton
USA • 2018
www.universal-publishers.com

978-1-59942-808-6 (pbk.)
978-1-59942-806-2 (ebk.)

Graphics by Jason Yang and Ameek Bindra
Edited by Dr. Sam Liu, MD, PhD Ophthalmologist
Cover design by Ivan Popov
Typeset by Medlar Publishing Solutions Pvt Ltd, India

Publisher's Cataloging-in-Publication Data

Names: Yang, Jason, author. | Pidgeon, Charles, author. | Bindra, Sehej, author. |
    Rastogi, Arnav, author. | Banerjee, Tirtho, author. | Gupta, Yash, author.
Title: Vision facts : questions about the human eye / Jason Yang, Charles Pidgeon,
    Sehej Bindra, Arnav Rastogi, Tirtho Banerjee, [and] Yash Gupta.
Description: Irvine, CA : Universal Publishers, 2018.
Identifiers: LCCN 2018934147 | ISBN 978-1-59942-808-6 (pbk.) |
    ISBN 978-1-59942-806-2 (ebook)
Subjects: LCSH: Vision--Popular works. | Eye--Miscellanea. | Visual perception. |
    Visual pathways. | Vision disorders. | Neurotransmitters. | BISAC: HEALTH &
    FITNESS / Vision. | REFERENCE / Questions & Answers.
Classification: LCC BF241 .Y36 2018 (print) | LCC BF241 (ebook) |
    DDC 152.14--dc23.

Dedicated
to
General Wei

Table of Contents

Questions

Answers

Preface

Teaching the science of vision is usually done in parts. For instance, eye anatomy may be taught alone in a class. Other classes may discuss the brain regions containing the distributed neural networks that process the photons registered in the retina. It is very difficult to obtain a more global view of vision because of such separation into parts. This book attempts to ask and answer questions from all parts of the visual system and is intended to be a supplement or reference book for anyone interested in vision who is involved with any education for vision. The book is also for the scientifically curious who want to know how vision works. Using this book will lead people to say with understanding, "You don't see with your eyes, you see with your brain".

This book is designed to merely ask a vision question, and provide an answer. If you are interested in the question, you can read the answer and follow up with a reference provided in the text. Below is a sample of the vision questions provided in this book.

A Few Examples

What is the resolution of the eye?
Why do astronomers often cover flashlights with a red filter while stargazing?
How many distinct colors are distinguishable by the human eye?
What is color-blindness and how does it occur?
What are the types of color-blindness?
Where do the retinal ganglion cells synapse in the brain?
What role does vision play in the circadian rhythm?
What is the function of each lobe of the cortex?

What are afterimages?
What is color constancy?
How does the eye develop?
What chemical signals cause the eye to develop into its mature form?
How does alcohol during pregnancy affect eye development?
How well can a newborn baby see?
How does being born prematurely affect vision?
How does color vision develop in infants?
What is the effect of visual deprivation in infants?
What is depth perception?
How do 3D glasses work?
What is 20/20 vision?
How does aging affect vision?

Introduction

Vision begins when a photon of light hits the surface of the eye. The photon first passes through the cornea and lens, which bend the path of the light to produce a sharp focused image. Most photons are focused onto the fovea, which is responsible for central vision and contains the highest ratio of cones to rods. Rods and cones are the two types of light-sensitive cells found in the retina. Rods are very sensitive but cannot distinguish color. This makes rods very useful for dim lighting and distinguishing brightness of an object. Cones help figure out the color of what you are viewing. Since there is a high number of cones in the fovea, the central portion of one's field of view is more vibrant than the peripheral vision. Since the lens is biconvex, the image presented on the retina is upside down.

When the photon reaches a rod, the chromophore of rhodopsin, 11-cis-retinal, is excited and changes configurations into all-trans-retinal. This molecular change on a rod triggers the splitting of rhodopsin and retinal. The freed opsin activates transducin, causing G protein based signal transduction. Eventually an electrical impulse is created and propagated. reaching the presynaptic membrane. At this point, calcium ions rush into the cell, causing the release of synaptic vesicles containing glutamate. A similar process occurs in cone cells using other opsins with retinal. Glutamate, the main excitatory neurotransmitter of the central nervous system, excites either a bipolar or horizontal cell which function to integrate information from multiple photoreceptor cells. Through direct or indirect (via inhibitory amacrine cells) means, the signal then reaches the ganglion cells. Multiple ganglion cell axons combine together to form the optic nerve, which leaves the retina at the optic disc. In this location, there are no photo-receptors which is why the optic disc is also known as the blind spot.

As the impulse travels through the optic nerve, it reaches a point known as the optic chiasm, where the optic nerve from each eye crosses over in an X-shape. At this point, the information is sorted such that the left portion of each retina's field of view (which is pro-jected onto the right half of each retina) is sent to the right optic tract, while the right portion of each retina's visual field is sent to the left optic tract. The majority of these fibers then head to the LGN (Lateral Geniculate Nucleus) of the thalamus, an important part of the brain that sorts out most sensory information and helps determine where it goes. From here the messages are relayed to higher order neurons in various regions of the brain. Other fibers of the optic tract go to the suprachiasmatic nucleus, superior colliculus and pretectum to func-tion in normalizing circadian rhythms, saccades and the pupillary light reflex respectively. The information at the LGN is separated into 6 layers. Layers 1 and 2 contain magnocellular cells, named for their large size, and Layers 3, 4, 5 and 6 contain parvocellular cells which are named for their small size. In between each of these layers there are koniocellular cells which are smaller than parvocellular cells. In addition to the size differences in these cells, there is also a difference in functionality. Magnocellular cells receive informations from rods which are necessary for movement, depth and visual acuity. Parvo-cellular cells receive information from L (red) and M (green) cones, and koniocellular cells receive information from S (blue) cones. All of these cells then travel via optic radiations to the primary visual cortex (V1). From here, the information travels along one of two pathways, the "What" pathway (Ventral Stream) and the "Where" pathway (Dorsal Stream). Along both of these pathways, the visual informa-tion stops at many important visual areas including V2, V3, V4, and V5. The difference in these pathways lies anatomically in the ending and functionally throughout the whole pathway. The ventral stream ends in the inferior temporal lobe and functions in perception, visual memory, and facial, object and pattern recognition. The dorsal stream ends in the posterior parietal lobe and functions in integrating visual stimuli with skilled movements, and motion tracking.

Acknowledgement

The authors would like to acknowledge the editor, Dr. Samuel M. Liu, MD PhD, an ophthalmologist, for both his conceptual and editorial corrections.

The authors would also like to acknowledge Ashay Bhatwadekar, PhD at the Eugene and Marilyn Glick Institute at the IU School of Medicine for his careful reading of the book.

Questions

Part I
The Visual Pathway

1 The Eye

2 Detection

3 Retinal Circuitry

4 Subcortical Structures

5 Cortical Circuitry

Q116 What are "interblobs"?

Q117 Who are David Hubel and Torsten Wiesel?

Q118 What are ocular dominance columns?

Q119 What are orientation columns?

Q120 How are blobs, ocular dominance columns, and orientation columns spatially arranged relative to each other?

Q121 What are higher-order visual areas?

Q122 What is V2?

Q123 If cytochrome oxidase staining in V1 forms blobs, what does the same stain show in V2?

Q124 What are "thin stripes"?

Q125 What are "thick stripes"?

Q126 What is retinal disparity?

Q127 What are "inter-stripes"?

Q128 What are illusory contours?

Q129 What is border ownership?

Q130 What is figure-ground segregation?

Q131 What is V3?

Q132 What is V4?

Q133 Is V4 the "color area"?

Q134 What are "globs"?

Q135 What are "interglobs"?

Q136 What is MT?

Q137 What is the two-streams hypothesis?

Q138 What is the dorsal stream?

Q139 What is the ventral stream?

Part II
Development

1 Fetal Development

2 Newborns and Infants

3 Children

Part III
Aging

Answers

Part I
The Visual Pathway

1 The Eye

Q1 **What are the important parts of the eye involved in detecting light?**

A1 Some major parts of the eye are the cornea, iris, pupil, lens, retina, and optic nerve.

"Parts of the Eye." *Rochester Institute of Technology.* Accessed September 08, 2017. https://www.cis.rit.edu/people/faculty/montag/vandplite/pages/chap_8/ch8p3.html.

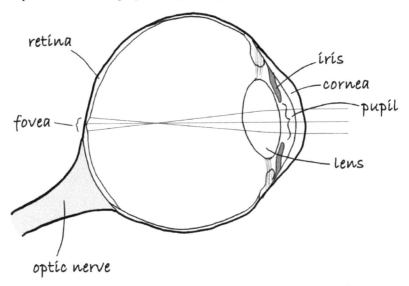

The eye is composed of various parts, including the cornea, iris, pupil, lens, retina, fovea, and optic nerve.

Q2 What is the cornea?

A2 The **cornea** is the outermost layer of the eye. It acts as a barrier against dirt, germs, and other harmful foreign particles and plays a key role in focusing light onto the retina.

   The cornea accounts for 65–75% of the bending of the light as it travels through the eye, known as refractive power. This process, called refraction, causes the image to focus on the retina at the back of the eye, as well as causing the image to be inverted horizontally and vertically. The sharpness of the image formed is known as **visual acuity**. Often, blurred vision is due to imperfections in the cornea.

   "Facts About the Cornea and Corneal Disease." *National Eye Institute*. May 01, 2016. Accessed September 08, 2017. https://nei.nih.gov/health/cornealdisease.

Q3 What is the iris?

A3 The **iris** is a circular muscle that surrounds the pupil. It can be dilated and constricted to regulate the amount of light entering the eyes by using the sphincter muscles to constrict the pupil and dilator muscles to dilate it. The iris also contains melanin that gives people their eye color.

   Daugman, J. "Anatomy, Physiology, and Development of the Iris." *University of Cambridge*. Accessed September 08, 2017. http://www.cl.cam.ac.uk/~jgd1000/anatomy.html.

Q4 What causes eye color?

A4 The pigmentation of the eye varies from light brown to black depending on the amount of deposition of pigments called **melanin** in the iris pigment epithelium and iris stroma. Blue, green, and hazel colored eyes result from phenomena known as Tyndall and **Rayleigh scattering** of light in the stroma, not because there is a blue or green pigment in the iris. The Rayleigh effect is also the reason the sky appears blue.

   "Is eye color determined by genetics?." *U.S. National Library of Medicine*. May 2015. Accessed September 08, 2017. https://ghr.nlm.nih.gov/primer/traits/eyecolor.

Q5 What is the pupil?

A5 The **pupil** is a hole at the center of the iris through which light enters the eye.

   "The Anatomy of the Eye." *The Physics Classroom*. Accessed September 08, 2017. http://www.physicsclassroom.com/class/refrn/Lesson-6/The-Anatomy-of-the-Eye.

Q6 **What is the lens?**

A6 The **lens** is a crystalline structure with an ellipsoid biconvex shape, a refractive power of about 18 **diopters**. It focuses light rays into the retina in order to create a clear image of objects at various distances through a process known as accommodation. The whole reflex is controlled by the cranial nerves II and III. The lens is held in place by fibers known as the **Zonule of Zinn**, also known as zonular fibers.

"The Anatomy of the Eye." *The Physics Classroom.* Accessed September 08, 2017. http://www.physicsclassroom.com/class/refrn/Lesson-6/The-Anatomy-of-the-Eye.

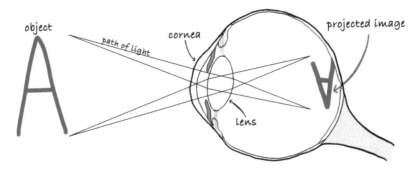

Images are inverted horizontally and vertically as they pass through the eye. The upper half of the visual field lands on the lower half of the retina, while the right half of the visual field lands on the left half of the retina.

Q7 **What is the ciliary body?**

A7 The **ciliary body** is a structure behind the iris that consists of the ciliary muscle and the ciliary epithelium. The **ciliary muscle** is involved in accommodation and changing the curvature of the lens. The **ciliary epithelium**, the other part of the ciliary body, is the structure in the eye that produces a fluid called the aqueous humor, which fills the anterior section of the eye.

"Ciliary body." *MedlinePlus.* August 14, 2015. Accessed September 08, 2017. https://medlineplus.gov/ency/imagepages/9188.htm.

Q8 **What is the ciliary muscle and what is its role in vision?**

A8 The **ciliary muscle** is a muscle around the lens that allows the eye to focus on objects of varying depth. There are several hypotheses describing this mechanism, which is known as **accommodation**. The **Helmholtz hypothesis**, described in 1855 by Hermann von

Helmholtz, states that the ciliary muscle pulls on the zonular fibers when relaxed. This tension on the zonular fibers pulls the lens into a flatter form. And when the ciliary muscle contracts, tension on the zonular fibers decreases, allowing the lens to become more round. On the other hand, the Schachar hypothesis proposes that the ciliary muscle does not pull on the zonular fibers when relaxed. When the ciliary muscle contracts, it directly pulls on the fibers, stretching the lens and causing it to become flatter.

In both hypotheses, the lens becomes flatter when viewing far objects and rounder when viewing near objects. This is because a flat lens has lower refractive power than a round lens, and thus bends incoming light less. Light rays coming from a distant object require less bending, as they are almost parallel.

"Theories of Accommodation." *American Academy of Ophthalmology*. Accessed September 08, 2017. https://www.aao.org/bcscsnippetdetail.aspx?id=f5f61688-98cd-4e30-84c0-b9acf775950c.

Q9 **What is the aqueous humor and what is its function?**

A9 The **aqueous humor** is a fluid produced by the ciliary epithelium that fills the anterior and posterior chambers. The **anterior chamber** is the region between the cornea and the iris, while the **posterior chamber** is the region between the iris and the zonule of Zinn. The functions of the aqueous humor include providing support for the eye as well as providing nutrients for structures that have no blood vessels, like the cornea and lens.

The pressure produced by the aqueous humor normally maintains the shape of the front of the eye, but becomes an issue in a disease known as glaucoma. When the aqueous humor fails to drain properly, the pressure increases and can damage sensitive neural tissue in the eye.

"Flow of Aqueous Humor." *BrightFocus Foundation*. June 21, 2017. Accessed September 09, 2017. http://www.brightfocus.org/glaucoma/infographic/flow-aqueous-humor.

Q10 **What is the vitreous humor and what is its function?**

A10 The **vitreous humor** is a clear, gel-like fluid occupying the space between the lens and the retina. This fluid is composed of mostly water, as well as collagen, sugar, and salts. Furthermore, the vitreous humor is stagnant, meaning it is not constantly being replenished, in contrast to the aqueous humor at the front of the eye. Therefore, if a foreign body enters the vitreous humor, it stays there, until it is surgically removed.

The function of the vitreous humor is to help transmit light to the retina and to maintain a certain amount of intraocular pressure to keep the layers of the retina tightly pressed together.

"Vitreous Humor Anatomy, Diagram & Function." *Healthline*. February 02, 2015. Accessed September 09, 2017. http://www.healthline.com/human-body-maps/vitreous-humor.

"Vitreous." *University of Minnesota*. Accessed September 20, 2017. http://teaching.pharmacy.umn.edu/courses/eyeAP/Eye_Anatomy/AssociatedStructures/Vitreous.htm.

Q11 **What is the optic nerve?**

A11 The **optic nerve** is formed from the convergence of axons from the retinal ganglion and leaves the eye from the optic disc. The **optic disc**, being the point of exit from the eye, has no overlying photoreceptors. Because of this, the optic disc is commonly referred to as the blindspot of the eye.

"The Optic Nerve (CN II) and Visual Pathway." *TeachMeAnatomy*. May 07, 2017. Accessed September 09, 2017. http://teachmeanatomy.info/head/cranial-nerves/optic-cnii/.

A demonstration of the blind spot: using your right eye, look at the black circle (top left). Then, slowly move your head forward and backward, while remaining fixed on the black circle, until the black square (top right) disappears. At this position, the visual information from the black square is projected onto the blind spot, which has no photoreceptors. As a result, the brain fills in the hole with surrounding information, which is white. To demonstrate this, repeat the procedure, but this time with the white circle (bottom left). At the correct position, the white square (bottom right) will turn black and disappear.

Q12   What is the retina?

A12 The **retina** is a part of the eye composed of neural tissue that covers around two-thirds of the back of the eye. It is responsible for the detection of light.

"Retina." *Encyclopædia Britannica.* April 28, 2017. Accessed September 09, 2017. https://www.britannica.com/science/retina.

Q13   What is the fovea?

A13 The **fovea centralis** is an area in the retina of your eye, that contains cones closely packed together. It is located in the center of the macula. The fovea allows for sharp central vision, which is imperative for many tasks such as driving. In fact, you are using your fovea right now to read this text.

"The Retina." *HyperPhysics.* Accessed September 09, 2017. http://hyperphysics.phy-astr.gsu.edu/hbase/vision/retina.html.

Q14   What is the resolution of the human eye?

A14 The resolution of the eye varies greatly and is highest at the fovea. The angular resolution at the fovea is about one arcminute, or 1/60th of a degree.

"Angular Resolution" [PowerPoint]. *University of Maryland.* Accessed September 09, 2017. https://www.astro.umd.edu/~thuard/astr288c/lecture6-notes.pdf.

Q15   What is the macula?

A15 The **macula** (also called macula lutea) is a yellowish area located near the center of the retina that contains the fovea.

"Macula." *American Academy of Ophthalmology.* March 07, 2017. Accessed September 09, 2017. https://www.aao.org/eye-health/anatomy/macula-6.

Q16   Why is the macula yellow?

A16 The macula is yellow due to the pigments lutein and zeaxanthin present in the area. Lutein and **zeaxanthin** are a type of carotenoid. **Carotenoids** are a class of pigments that include yellow, orange, and red. The macula contains a majority of zeaxanthin, while lutein is usually in other parts of the retina. These pigments are thought to prevent the macula from degeneration by absorbing high energy blue and ultraviolet light that could damage the retina.

"Lutein & Zeaxanthin." *American Optometric Association.* Accessed September 09, 2017. https://www.aoa.org/patients-and-public/caring-for-your-vision/diet-and-nutrition/lutein?sso=y.

"Carotenoids." *Linus Pauling Institute.* February 13, 2017. Accessed September 09, 2017. http://lpi.oregonstate.edu/mic/dietary-factors/phytochemicals/carotenoids.

Q17 What is the tapetum lucidum?

A17 The **tapetum lucidum** is a layer that is located directly behind the retina. It is found in certain nocturnal animals that require sensitive night vision, such as cats and dogs. Humans lack a tapetum lucidum. The tapetum lucidum is highly reflective to increase the amount of light available to the retinal cells for detection. However, by reflecting light around the eye, the tapetum lucidum also blurs vision, reducing visual acuity.

Ollivier FJ, Samuelson DA, Brooks DE, Lewis PA, Kallberg ME, Komáromy AM. "Comparative morphology of the tapetum lucidum (among selected species)." *Vet Ophthalmol.* 2004 Jan-Feb;7(1):11–22. https://www.ncbi.nlm.nih.gov/pubmed/?term=14738502.

Q18 What is the retinal pigment epithelium?

A18 The **retinal pigment epithelium** (RPE) is a layer of cells just outside the retina. The function of the RPE is to nourish the retinal cells and to participate in the regeneration of the photopigments in the retina. Unlike the tapetum lucidum, the RPE functions to absorb light. This increases visual acuity by preventing light from reflecting around inside the eye. Absorption of excess light also helps prevent damage to the retinal cells from high light intensities.

Purves, D., G. J. Augustine, and D. Fitzpatrick. "The Retina." In *Neuroscience.* 2nd ed. Sunderland, MA: Sinauer, 2001. Accessed September 20, 2017. https://www.ncbi.nlm.nih.gov/books/NBK10885/.

Lamb, T. D., & Pugh, E. N., Jr. (2006). "Phototransduction, Dark Adaptation, and Rhodopsin Regeneration The Proctor Lecture." *Invest. Ophthalmol. Vis. Sci.,* 47(12). doi:10.1167/iovs.06-0849.

Q19 What is the choroid?

A19 The **choroid** is a structure that lies between the sclera and the retina, and is the posterior part of the uvea, which consists of the choroid, iris, and ciliary body. It is a vascular tissue that provides all layers of the eye with approximately 90% of their blood supply, which it receives from the central retinal artery.

"Choroid." *University of Minnesota.* Accessed September 20, 2017. http://teaching. pharmacy.umn.edu/courses/eyeAP/Eye_Anatomy/CoatsoftheEye/Choroid.htm.

Q20   What is the sclera?

A20   The **sclera** forms the surrounding wall of the eyeball and is the white portion of the eye. The sclera is continuous with the cornea and is covered by a clear mucus known as the conjunctiva.

The sclera is a form of tough connective tissue containing **collagen** embedded in extra fibrillar matrix composed of proteoglycans. The sturdy composition of the sclera helps it to maintain pressure in the eye, and withstanding the forces of the extraocular muscles. Furthermore, the sclera provides support for structure such as the retina, and maintains eye shape during movement. The sclera also contains fluid outflow channels to maintain intraocular pressure.

Meek, K. M. "The Cornea and Sclera." In *Collagen*, 359–96. Boston, MA: Springer, 2008. Accessed September 20, 2017. doi:10.1007/978-0-387-73906-9_13.

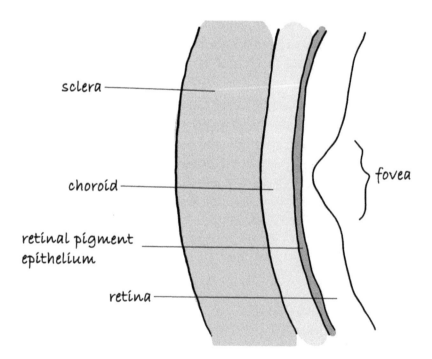

The layers of the back of the eye (from inside to outside) include the retina, retinal pigment epithelium, choroid, and sclera. The fovea is a spot on the center of the retina responsible for central vision.

2 Detection

Q21 **What is sensory transduction?**

A21 **Sensory transduction** is the process where energy from the surrounding environment is converted to electrical impulses by receptors. In the case of vision, the receptors are represented by the photoreceptor cells of the retina.

"Sensory transduction." *National Center for Biotechnology Information.* Accessed September 09, 2017. https://www.ncbi.nlm.nih.gov/books/NBK10981/def-item/A2859/.

Q22 **What types of neurons are found in the retina?**

A22 In the retina, there are five different types of neurons: photoreceptor cells, bipolar cells, amacrine cells, horizontal cells, and retinal ganglion cells. These neurons are stacked in three layers inside the retina.

Purves, D., G. J. Augustine, and D. Fitzpatrick. "The Retina." In *Neuroscience.* 2nd ed. Sunderland, MA: Sinauer, 2001. Accessed September 20, 2017. https://www.ncbi.nlm.nih.gov/books/NBK10885/.

Q23 **What are the layers of the retina?**

A23 There are three cell layers in the retina. Cell bodies are located in the **inner nuclear**, **outer nuclear**, and **ganglion cell layers**. There are two additional layers containing the synapses between cell layers; these layers are known as the **inner plexiform** and **outer plexiform** layers.

Purves, D., G. J. Augustine, and D. Fitzpatrick. "The Retina." In *Neuroscience.* 2nd ed. Sunderland, MA: Sinauer, 2001. Accessed September 20, 2017. https://www.ncbi.nlm.nih.gov/books/NBK10885/.

Q24 **How do the layers of the retina work together?**

A24 First, a photon of light is detected by the photoreceptor cell. This initiates a sequence of events that changes the membrane potential of the photoreceptor cell. The change in membrane potential causes a decrease in the amount of neurotransmitter released to bipolar cells. The bipolar cells then transmit the information to ganglion cells, whereby the axons form the optic nerve carries the information to the brain.

Purves, D., G. J. Augustine, and D. Fitzpatrick. "The Retina." In *Neuroscience.* 2nd ed. Sunderland, MA: Sinauer, 2001. Accessed September 20, 2017. https://www.ncbi.nlm.nih.gov/books/NBK10885/.

Q25 **What are photoreceptor cells?**

A25 **Photoreceptor cells**, also known as photosensitive cells, are specialized neurons that are located in the retina. These cells convert light into chemical and electrical signals which are then passed to the other layers of the retina, the optic nerve, and then finally to the brain for processing.

"Photoreceptor Cells." *National Center for Biotechnology Information.* Accessed September 09, 2017. https://www.ncbi.nlm.nih.gov/pubmedhealth/PMHT0024257/.

Q26 **What are the two types of photoreceptors?**

A26 The retina contains rod-shaped and cone-shaped photoreceptor cells (**rods** and **cones**). Both rods and cones have three parts: an outer segment, an inner segment, and the synaptic ending. The outer segment, consisting of many **membranous discs**, is shaped as either a rod or a cone, giving the photoreceptors their name. The inner segment consists of the cell body, including the nucleus and the **mitochondria**. And finally the synaptic terminals, that are connected to the inner body, contact either bipolar or horizontal cells.

"The Retina." *Neuroscience For Kids.* Accessed September 09, 2017. https://faculty.washington.edu/chudler/retina.html.

Q27 **What is the difference between the rods and cones?**

A27 There is only one type of rod cell, and so based solely on information from rods, the color of light cannot be determined. However, rods alone are able to determine the luminance, or brightness, of light. On the other hand, there are three types of cones. With these cones, the chrominance, or color, of the light can also be determined. Cones are not without disadvantages though, as a single cone is much less sensitive to light than a single rod. As a result, rods become more important than cones in low-light conditions. However, it is important to note that the relationship between rods and cones is much more complicated than this, as both rods and cones contribute to both the midget and the parasol systems. Beyond the retina, there is no distinction between the "rod system" and "cone system", as inputs from both types of photoreceptors are combined into the various pathways that will be discussed later.

"Rods and Cones." *Rochester Institute of Technology.* Accessed September 09, 2017. http://www.cis.rit.edu/people/faculty/montag/vandplite/pages/chap_9/ch9p1.html.

"Rods and Cones." *HyperPhysics.* Accessed September 09, 2017. http://hyperphysics.phy-astr.gsu.edu/hbase/vision/rodcone.html.

Q28 **How many rods are there in an average human retina?**

A28 There are estimated to be over 120 million rods in the average human retina.

"Rods and Cones." *HyperPhysics*. Accessed September 09, 2017. http://hyperphysics. phy-astr.gsu.edu/hbase/vision/rodcone.html.

Q29 **How many cones are there in an average human retina?**

A29 There are somewhere between 6 to 7 million cones in the average human retina, of which 65% detect red light, 33% detect green light, and only 2% detect blue light.

"Rods and Cones." *HyperPhysics*. Accessed September 09, 2017. http://hyperphysics. phy-astr.gsu.edu/hbase/vision/rodcone.html.

Q30 **What is the distribution of rods and cones throughout the retina?**

A30 The number of cones is greatest at the fovea. Farther from the fovea, the number cones decreases. On the other hand, there are few rods in the fovea. The majority of the rods are located in the periphery of the retina. As a result, in low-light conditions the center of the visual field is not the most sensitive, but rather the area surrounding it.

"Rods and Cones." *HyperPhysics*. Accessed September 09, 2017. http://hyperphysics. phy-astr.gsu.edu/hbase/vision/rodcone.html.

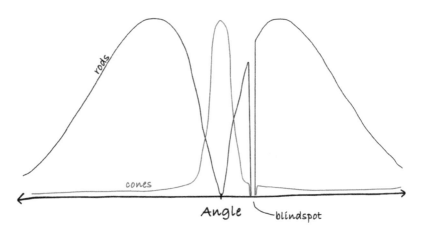

The graph shows the density of rods and cones. The number of cones is highest at the fovea and decreases as you get farther from the center. The number of rods is highest surrounding the fovea, but decreases toward the center and toward the periphery, with no rods located in the fovea. The blind spot is also marked, where there are no rods or cones.

Q31 **What is the structure of a photoreceptor cell?**

A31 A photoreceptor consists of the outer segment, inner segment, cell body, and synaptic terminal. The **outer segment** contains membranous discs, embedded with either rhodopsin or photopsin proteins that detect light. The discs are stacked throughout the outer segment. The **inner segment** contains other organelles required by the photoreceptor, such as the ribosomes and mitochondria. The cell body contains the nucleus and the genome. The synaptic terminal is the junction between the photoreceptor and other neurons.

Kolb, H. (n.d.). "Photoreceptors by Helga Kolb." *Webvision*. Retrieved from http://webvision.med.utah.edu/book/part-ii-anatomy-and-physiology-of-the-retina/photoreceptors/.

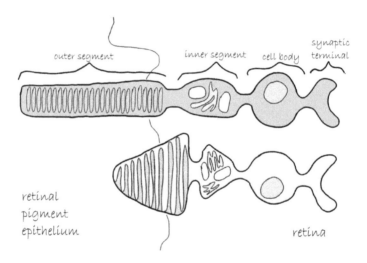

The rods and cones consist of an outer segment, inner segment, cell body, and synaptic terminal. The outer segment is located on the border between the retina and the retinal pigment epithelium.

Q32 **What is rhodopsin?**

A32 Rhodopsin is a type of **G-protein coupled receptor** (GPCR), which when activated, causes a decrease in the levels of a chemical called **cAMP** in the cell. Rhodopsin belongs to a class of proteins called **opsins**, which are proteins responsible for detection of light found in photoreceptors.

Rogers, Kara. "Rhodopsin." *Encyclopædia Britannica*. Accessed September 21, 2017. https://www.britannica.com/science/rhodopsin.

Q33 How do rods detect light?

A33 Rods use **rhodopsin**, which consist of a protein attached to a chemical called **retinal** (which comes from vitamin A). Normally the retinal is in the form of 11-cis-retinal, but when it absorbs light, it becomes trans-retinal. This change in structure of retinal activates the rhodopsin molecule and thus detects light.

"Rods & Cones." *Rochester Institute of Technology.* Accessed September 21, 2017. https://www.cis.rit.edu/people/faculty/montag/vandplite/pages/chap_9/ch9p1.html.

Q34 What happens when a rhodopsin molecule becomes activated?

A34 In a normal unactivated photoreceptor, sodium is constantly entering the cell, causing a neurotransmitter called glutamate to be released. The decrease in cAMP levels causes sodium to stop entering the cell, preventing the further release of glutamate. This decrease in glutamate levels signals the detection of light.

Rogers, Kara. "Rhodopsin." *Encyclopædia Britannica.* Accessed September 21, 2017. https://www.britannica.com/science/rhodopsin.

Q35 What is the absorption spectrum of rhodopsin?

A35 The absorption spectrum of rhodopsin peaks at a wavelength of about 500 nanometers.

Berg, J. M., J. L. Tymoczko, and L. Stryer. "Photoreceptor Molecules in the Eye Detect Visible Light." In *Biochemistry.* 5th ed. New York: W. H. Freeman, 2002. Accessed September 21, 2017. https://www.ncbi.nlm.nih.gov/books/NBK22541/.

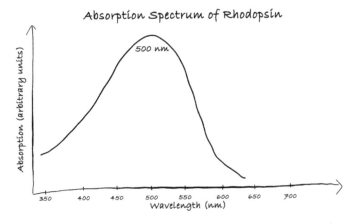

The absorption spectrum of rhodopsin shows the amount of light absorbed as a function of the wavelength of the light. Peak absorption occurs with light of wavelength 500 nanometers.

Q36 What is the Purkinje effect and why does it occur?

A36 The **Purkinje effect** was described by Jan Evangelista Purkinje. He noticed that the green leaves and red flowers of a plant seemed equally bright under photopic, or bright, conditions. However, under scotopic, or dim, conditions, such as at night, the red flowers seemed far less bright than the green leaves. This is due to the absorption spectrum of rhodopsin. Rods are the primary photoreceptors used in scotopic conditions, so due to the low sensitivity of rhodopsin in the red wavelengths, the red flower seems less bright.

"Purkinje Effect." *Encyclopædia Britannica.* Accessed September 21, 2017. https://www.britannica.com/topic/Purkinje-effect.

Q37 What is photobleaching?

A37 When 11-cis-retinal absorbs light, it undergoes a conformational change into all-trans-retinal, signalling the detection of light. However, all-trans-retinal cannot detect light anymore. When exposed to bright lights, rod and cone cells **photobleach**, or become less sensitive to light, as more 11-cis-retinal is converted to all-trans-retinal.

On the other hand, when in a dark environment for an extended period of time, the all-trans-retinal is eventually converted back to 11-cis-retinal through a series of enzymatic reactions. This process, known as **dark adaptation**, allows the eye to be more sensitive to light under dim conditions, such as at night.

"photobleaching." *Oxford University Press.* Accessed September 21, 2017. https://en.oxforddictionaries.com/definition/photobleaching.

Q38 Why do astronomers often cover flashlights with a red filter while stargazing?

A38 Astronomers use red flashlights to maintain dark adaptation for viewing during the night. When rods are exposed to bright light, even for a brief moment, they become photobleached, and therefore become less sensitive to light for around half an hour, until more 11-cis-retinal can be regenerated. However, since the absorption spectrum of rods does not cover long wavelengths, they photobleach minimally to red light. As a result, the cones are used to view the red light while maintaining the sensitive rod cells for viewing faint stars.

West, Jim. "Red Lights for Astronomy—Skilight Mini." *Colorado Springs Astronomical Society*. Accessed September 21, 2017. http://csastro.org/red-lights-for-astronomy-skilight-mini/.

Q39   How do cones detect light?

A39   Cones have a similar process to rods for detecting light except use a protein called **photopsin** instead of rhodopsin on their discs. There are 3 types of cones, each with a distinct type of photopsin, which are sensitive to a different range of wavelengths of light: red, green, and blue (or L, M, and S for long, medium, and short wavelength). Together, cones allow use us to view different colors of light.

"Rods and cones of the human eye." *Arizona State University*. Accessed September 21, 2017. https://askabiologist.asu.edu/rods-and-cones.

Q40   What are the absorption spectra of the three types of photopsin?

A40   β photopsin found in **S cones**, has an absorption spectrum covering 400–500 nm (blue); γ photopsin, in **M cones**, covers 450–630 nm (green); ρ Photopsin, in **L cones**, has an absorption spectrum of 500–700 nm (red).

Purves, D., G. J. Augustine, and D. Fitzpatrick. "The Retina." In *Neuroscience*. 2nd ed. Sunderland, MA: Sinauer, 2001. Accessed September 20, 2017. https://www.ncbi.nlm.nih.gov/books/NBK10885/.

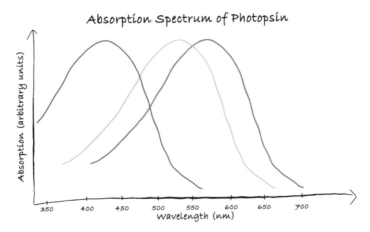

The absorption spectra of photopsins show the amount of light absorbed as a function of the wavelength of the light. Peak absorption occurs in the blue, green, and red regions of the visual spectrum for S-, M-, and L-cones, respectively.

Q41 **Why is vitamin A important for the proper functioning of the eye?**

A41 **Vitamin A** is specifically important for the functioning of the retina, as 11-cis-retinal is an isomer derived from vitamin A, and the absence of 11-cis-retinal leads to the absence of functioning rhodopsin and photopsin, which can lead to night blindness.

Boyd, Kierstan. "What Is Vitamin A Deficiency?" *American Academy of Ophthalmology.* November 8, 2012. https://www.aao.org/eye-health/diseases/vitamin-deficiency.

Higdon, Jane. "Vitamin A." *Oregon State University.* 2000. http://lpi.oregonstate.edu/mic/vitamins/vitamin-A.

"11-cis-Retinal." *National Institute of Health.* https://pubchem.ncbi.nlm.nih.gov/compound/11-cis-Retinal.

"Retinal." *Imperial College London.* https://www.ch.ic.ac.uk/vchemlib/mim/bristol/retinal/retinal_text.htm.

Q42 **What is disc shedding?**

A42 In photoreceptors, the photopigments (e.g. rhodopsin) are packed in membranous discs. The discs are formed from near the inner segment of the photoreceptor and grows towards the the tip of the outer segment, where they are shed, similar to how skin grows outwards and sheds off once it reaches the surface. Through this process, old photobleached pigments are shed and replaced by new photopigments.

Kocaoglu, Omer P., Zhuolin Liu, Furu Zhang, Kazuhiro Kurokawa, Ravi S. Jonnal, and Donald T. Miller. "Photoreceptor disc shedding in the living human eye." *Biomed. Opt. Express* 7, no. 11 (2016): 4554. doi:10.1364/boe.7.004554.

Q43 **Why must light rays pass through the non-photosensitive elements of the retina before reaching the photoreceptor cells?**

A43 The pigment epithelium is responsible for removing the shedded discs from photoreceptors. Additionally, the pigment epithelium is able to regenerate new photopigment molecules. Therefore, the pigment epithelium must be located adjacent to the outer segment of photoreceptor cells. For the photoreceptor cells to be adjacent to the epithelium, they must be located in the outermost region of the retina.

Strauss, Olaf. "The retinal pigment epithelium." *University of Utah.* http://webvision.med.utah.edu/book/part-ii-anatomy-and-physiology-of-the-retina/the-retinal-pigment-epithelium/.

Q44 **How many distinct colors are distinguishable by the human eye?**

A44 Various papers estimate this figure to be between 100,000 and 10 million different colors.

"How many colors can we see?" *Washington State University.* July 13, 2016. https://askdruniverse.wsu.edu/2016/07/13/many-colors-can-see/.

Q45 **How can only three types of cone cells detect possibly up to 10 million different colors?**

A45 The cones can distinguish a huge variety of different colors through using different ratios of the primary colors detected by the individual cone cells. It was evolutionary and efficient to produce human color vision using only three types of cells instead of one for each color of the whole visible spectrum that the human eyes can detect. For example, yellow light would stimulate both red and green cones, and purple light would stimulate both blue and red cones.

Keheller, Colm. "How we see color." *TED-Ed.* https://ed.ted.com/lessons/how-we-see-color-colm-kelleher#review.

Q46 **What is the Young-Helmholtz trichromatic theory?**

A46 The **Young-Helmholtz Theory** is based of on the works of Thomas Young and **Hermann von Helmholtz** in the 1800's. Young's theory states that there are only three types of photoreceptors to detect color, and each was sensitive to a particular range of the visible light spectrum.

Many years later, Hermann von Helmholtz developed the theory further saying that the three types of color photoreceptors were: short-preferring (blue), mid-preferring (green), and long-preferring (red). Other colors use different proportions of red blue and green, for example yellow light uses different proportions of red and green.

Heesen, Remco. The Young-(Helmholtz)-Maxwell Theory of Color Vision. Report. Philosophy, *Carnegie Mellon University.*

Q47 **What are the proportions of the three types of cones in the fovea?**

A47 The ratio of red:green:blue cones in the fovea is approximately 12:6:1. There are far fewer blue cones than there are red and green cones. This is due to an optical property known as **chromatic aberration,**

in which light rays of different wavelengths are refracted in different amounts. The peak wavelengths of red and green cones are relatively close, while the peak wavelength of blue cones is significantly different. As a result, the light detected by red and green cones can be in focus, but blue light is refracted more and is out of focus. Thus, red and green cones dominate the fovea, where an accurate focus is necessary for high acuity vision.

Fairchild, Mark D. *Color Appearance Models.* Wiley, 2013. doi:10.1002/9781118653128.

Gouras, Peter. "Color Vision." *University of Utah.* July 1, 2009. http://webvision.med.utah.edu/book/part-vii-color-vision/color-vision/.

Q48   What is color-blindness and how does it occur?

A48   Color blindness is a condition characterized by a decreased ability to distinguish colors. It occurs when one or more types of cone cells are either absent, nonfunctioning, or detect different colors than they are suppose to. Color blindness ranges from mild, where one cone cell functions abnormally, to severe color blindness where all cone cells function abnormally.

"What Is Color Blindness?" *American Academy of Ophthalmology.* December 11, 2013. https://www.aao.org/eye-health/diseases/what-is-color-blindness.

Q49   What are the types of color-blindness?

A49   People who have malfunctioning cones are called anomalous trichromats. Trichromats are people who have normal vision and all three of their cones are properly functioning. However in anomalous trichromats, one type of cone malfunctions. The malfunctioning cone contains photopsin with a different absorption spectrum than normal. There are three types of anomalous trichromacies: **protanomaly**, **deuteranomaly**, and **tritanomaly**. In protanomaly, the absorption spectrum of the red photopsin is shifted toward the green photopsin, so the ability to distinguish between red and green colors is diminished. In deuteranomaly, the absorption spectrum of green photopsin is shifted toward red photopsin, again diminishing red-green color discrimination. Finally, in tritanomaly, the blue photopsin malfunctions, diminishing blue-green color discrimination.

The effects of protanomaly, deuteranomaly, and tritanomaly also vary—ranging from close to normal perception of the faulty color to a complete absence of the color.

Red-green colorblindness includes people with both protanomaly and deuteranomaly. Not only do they have trouble distinguishing between reds and greens, but also confuse brown and oranges, and blue and purple. On the other hand, people who are less sensitive to blue light have difficulty between the differences of blue and yellow, violet and red, and blue and green. The world seems to be mainly red, pink, black, white, gray, and turquoise.

Dichromacy is a condition where someone has only two types of functional cones. This results in a lack of ability to a see a certain part of the color spectrum, whether it is red (protanopia), green (deuteranopia), or blue (tritanopia).

Finally, the last condition is known as **monochromacy**. People with this condition see the world like a black and white movie. The only color they can perceive are different shades of gray. Fortunately, this rare condition occurs only in 1/33,000 people.

"Types of Colour Blindness." *Colour Blind Awareness.* http://www.colourblind awareness.org/colour-blindness/types-of-colour-blindness/.

Q50 **What percentage of the people are color-blind? Who are at risk of this condition?**

A50 Gender plays an important role in determining color blindness, as both protanomaly and deuteranomaly are sex linked traits. Worldwide, 8% of men are colorblind, and only 0.5% of women are colorblind. Another factor is ethnicity, where Caucasians have a higher risk compared to Africans. Some countries have significantly higher prevalence of color blindness, including Brazil and India.

"Types of Colour Blindness." *Colour Blind Awareness.* http://www.colourblind awareness.org/colour-blindness/types-of-colour-blindness/.

Q51 **How is color-blindness diagnosed?**

A51 Color-blindness can be diagnosed using several different types of tests. One of these tests, the Ishihara test, uses multiple dots arranged in a circle. Each of these dots is a shade of one of two colors. Together, all of the dots of one color form a number. The test involves asking the patient to identify the number written on the plate. The complete Ishihara test uses 38 different plates. In some plates, only a person who is not color-blind can read the number on the plate. In other plates, only a person who is color-blind can identify the number. Finally, in a third type of plate, the number visible to a person with normal vision is different from the number visible to a person who is color-blind.

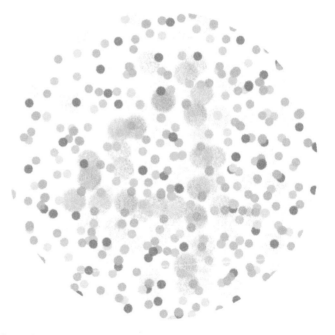

An Ishihara plate would look something like this. Note that this image is not intended to diagnose color-blindness.

The Farnsworth-Munsell 100 hue test is another type of color-blindness test used diagnostically. This test assesses the ability of the patient to distinguish minute differences in color. In the Farnsworth-Munsell 100 hue test, the patient is asked to arrange 100 different colored tiles in order based on their colors. According to a study done by Pacific University, the Ishihara test is generally more accurate and less time consuming than the Farnsworth Munsel 100 hue test.

Murphy, Rachel A. 2015. "Comparing Color Vision Testing Using the Farnsworth-Munsell 100-Hue, Ishihara Compatible, and Digital TCV Software." Master's thesis, Pacific University. Accessed September 21, 2017. http://commons.pacificu.edu/opt/9/.

3 Retinal Circuitry

Q52 **How does a neuron receive and transmit signals?**

A52 Neurons receive signals from nearby neurons through an extension that branches along the cell body known as a **dendrite**. The **synapse** is the small gap between two neurons, usually at a dendrite. These dendrites receive chemical signals, known as **neurotransmitters** that diffuse across the synapse and induce changes to the neuron. When a neuron receives the chemical signal through the dendrite, ions such as sodium rush into the cell causing a change in the charge of the inside of the neuron from negative to positive. The difference in charge between the inside of the cell and the fluid surrounding it is known as the **membrane potential**. Once the neuron's membrane potential changes, it creates an **action potential**, an impulse that travels along the cell's axon. The **axon** is a long extension of neuron that allows it to send signals across long distances, usually forming a synapse with another neuron.

"The Neuron." *Brain Facts*. April 1, 2012. http://www.brainfacts.org/brain-basics/neuroanatomy/articles/2012/the-neuron/.

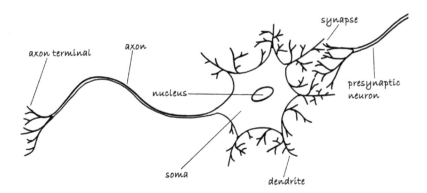

A typical multipolar neuron

A typical multipolar neuron consists of dendrites, a soma (cell body), a nucleus as well as various other organelles, an axon, and axon terminals. Multipolar refers to the fact that there are many processes extending from the neuron, including an axon and several dendrites. However, not all neurons are multipolar; neurons are extremely diverse in their morphologies. Other neuron classifications by shape include unipolar, pseudounipolar, and bipolar.

Q53 **What is a receptive field?**

A53 In vision, a **receptive field** of a cell is the area of the visual field in which the cell responds to stimulus. The **visual field** is the total area that is visible to a person at a given time, spanning about 200 degrees horizontally and 150 degrees vertically.

For example, consider neuron A in the brain responsible for some aspect of vision processing. Neuron A receives inputs from neurons B, C, and D, each of which receive inputs from several other neurons, and so on. If the inputs of neuron A were traced to the source, we would find a collection of photoreceptors, each of which detects light from a certain area of the visual field. When these areas of the visual field are combined, they form the receptive field of neuron A. When light from within the receptive field is present, a photoreceptor is activated which activates other neurons that eventually lead to the neuron A. Thus, a neuron responds to stimulus within its receptive field. On the other hand, if light is present but outside of the receptive field, a photoreceptor would still detect it, but the signal from this photoreceptor would not make its way to neuron A, but rather a different neuron. Thus, a neuron does not respond to stimulus outside of its receptive field.

Spector, Robert A., and Robert A. Spector. "Chapter 116." In *Clinical Methods: The History, Physical, and Laboratory Examinations*. 3rd ed.

Krantz, John H. "Receptive Field Tutorial." *Hanover College*. http://psych.hanover. edu/Krantz/receptive/.

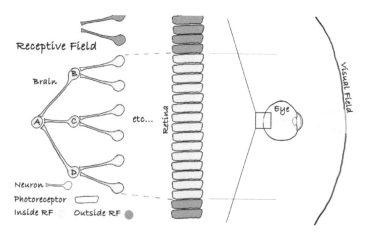

The receptive field of neuron A is the region of the visual field where the light is detected and the signal is ultimately sent to neuron A. In other words, neuron A responds to changes in visual stimuli within the receptive field, but not outside of it.

Q54 **How does receptive field size affect spatial acuity?**

A54 As receptive field of a ganglion cell increases, detail decreases because a larger number of photoreceptors send signals to a single ganglion cell, and thus one ganglion cell output represents a larger area.

Alonso, Jose-Manuel, and Yao Chen. "Receptive field." *Scholarpedia* 4, no. 1 (2009): 5393. doi:10.4249/scholarpedia.5393.

Hubel, David H. *Eye, Brain, and Vision*. Scientific American Library, 1988. Accessed September 21, 2017. http://hubel.med.harvard.edu/.

Q55 **How does the size of receptive fields vary throughout the retina?**

A55 Retinal ganglion cells in the fovea centralis have the smallest receptive fields, while the ganglion cells outside the fovea, the visual periphery, have larger receptive fields. This explains why the visual periphery has poor spatial resolution. For example, try reading this sentence while staying fixated on one word. This task is impossible for the human eye, for the word you are fixated on is being projected onto your fovea centralis, and the rest is being projected onto your peripheral retina.

Balasubramanian, Vijay, and Peter Sterling. "Receptive fields and functional architecture in the retina." *J. Physiol.* 587, no. 12 (2009): 2753–767. doi:10.1113/jphysiol.2009.170704.

Heeger, David. "Perception Lecture Notes: Retinal Ganglion Cells." *New York University.* http://www.cns.nyu.edu/~david/courses/perception/lecturenotes/ganglion/ganglion.html.

Q56 **What is center-surround antagonism?**

A56 **Center-surround antagonism** is the opposing interaction between the center and surrounding regions of a receptive field of certain neurons in the visual pathway, such as retinal ganglion cells. This process allows the eye to detect the edges of objects (edge detection) through sensing contrasts in brightness.

Center-surround receptive fields can be divided into two types: on-center and off-center. In a cell with an on-center receptive field, light falling on the central portion of the receptive field stimulates the cell, while light falling on the outer portion of the receptive field inhibits the cell. Thus, if the entire receptive field is illuminated, the cell only responds weakly, while if only the center is illuminated, the cell responds more strongly.

In off-center cells, the opposite occurs. Light falling on the central portion inhibits the cell, while light falling on the outer portion stimulates the cell. Thus, the cell responds strongest when only the outer portion of the receptive field is illuminated.

Note that although center-surround receptive fields are found in many cells in the retina, the cells in the brain exhibit more complex receptive fields for further processing of visual information.

Heeger, David. "Perception Lecture Notes: Retinal Ganglion Cells." *New York University.* http://www.cns.nyu.edu/~david/courses/perception/lecturenotes/ganglion/ganglion.html.

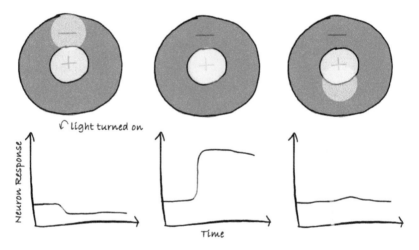

A typical center-surround receptive field consists of a center and a surround. The cell shown is on-center, as when light is shined on the center of the receptive field, the cell increases its response, and when light is shined on the surrounding region of the receptive field, the cell decreases its response. The opposite would occur in an off-center cell.

Q57 **What are bipolar cells?**

A57 **Bipolar cells** are neurons in the retina that provides a direct link from the photoreceptor cell to the ganglion cells. They synapse with the photoreceptors in the inner plexiform layer and with the ganglion cells in the outer plexiform layer.

"Bipolar cells." *National Institute of Health.* https://www.ncbi.nlm.nih.gov/books/NBK10981/def-item/A2306/.

Q58 **What are the types of bipolar cells?**

A58 There are eleven types of bipolar cells. Ten of these types of bipolar cells receive inputs from cones only, while one type of bipolar cells receives input from only rods. The separate channels of rod information and cone information later converge on retinal ganglion cells.

Euler, Thomas, Silke Haverkamp, Timm Schubert, and Tom Baden. "Retinal bipolar cells: elementary building blocks of vision." *Nat. Rev. Neurosci.* 15, no. 8 (2014): 507–19. doi:10.1038/nrn3783.

Ghosh, Krishna K., Sascha Bujan, Silke Haverkamp, Andreas Feigenspan, and Heinz Wässle. "Types of bipolar cells in the mouse retina." *J. Comp. Neurol.* 469, no. 1 (2003): 70–82. doi:10.1002/cne.10985.

Q59 **What are horizontal and amacrine cells?**

A59 Horizontal and amacrine cells are two types of neurons that are responsible for lateral interactions of the retina and the formation of receptive fields. Both cells have their cell body located in the inner nuclear layer. Horizontal cells and amacrine cells have different functions throughout the retina.

Purves, Dale. *Neuroscience.* 2nd ed. Sunderland, MA: Oxford University Press, 2001.

Q60 **What is the function of horizontal cells?**

A60 **Horizontal cells** are thought to provide negative feedback to the photoreceptors, allowing the eye to see well under bright and dim lighting. The activation of photoreceptors causes the activation of horizontal cells, which in turn inhibits all of the photoreceptors in the region. Thus, when the image is too bright, the eye adapts to the lighting condition by inhibiting photoreceptors.

Hubel, David H. *Eye, Brain, and Vision.* New York City, New York: Scientif. American Library, 1995.

Q61 **What is the function of amacrine cells?**

A61 **Amacrine cells** processes are located in the inner plexiform layer. They receive information from bipolar cells, and conduct impulses to ganglion cells. Each subclass of amacrine cells have different functions. For example, amacrine cells carry visual information from rod bipolar cells to retinal ganglion cells, as the two do not directly synapse.

Purves, D., G. J. Augustine, and D. Fitzpatrick. *Neuroscience*. 2nd ed. Sunderland, MA: Oxford University Press, 2001.

Q62   What are ganglion cells?

A62   Ganglion cells are another group of neurons located in the retina. The ganglion cells are the last cells in the retina that receives visual information, and are located in the innermost layer of the retina. These cells receive information from bipolar cells and amacrine cells that are also located in the retina.

"Retinal Ganglion Cell." *NeuronBank.* http://neuronbank.org/wiki/index.php/ Retinal_Ganglion_Cell.

Purves, D., G. J. Augustine, and D. Fitzpatrick. *Neuroscience*. 2nd ed. Sunderland, MA: Oxford University Press, 2001.

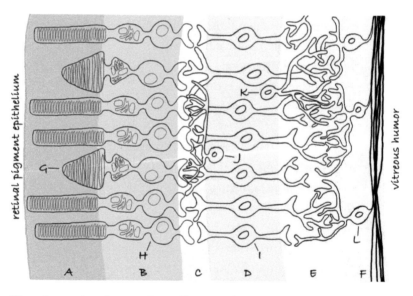

The retina consists of various layers (from outside to inside): A—photoreceptor layer, B—outer nuclear layer, C—outer plexiform layer, D—inner nuclear layer, E—inner plexiform layer, F—ganglion cell layer. Various cells are also labeled: G—cone, H—rod, I—bipolar cell, J—horizontal cell, K—amacrine cell, L—ganglion cell.

Q63   What are Mach bands?

A63   Mach bands are a type of illusion named after Ernst Mach. In the image, each rectangle appears lighter on the left side than on the right side, even though each is of uniform color. This phenomenon can be explained by the receptive fields of the ganglion cells in the eye. For neurons on the border between a darker and a lighter rectangle, the center falls on one rectangle while part of the inhibitory surround falls on the other rectangle.

"Mach Bands." *University of Calgary.* http://www.ucalgary.ca/pip369/mod3/brightness/machbands.

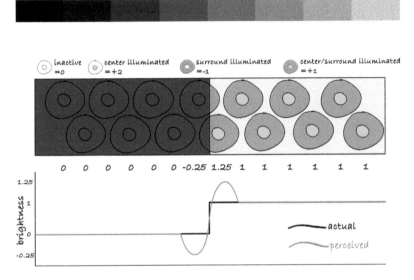

In the phenomenon of Mach bands, shown by the eight shaded rectangles, the left side of the rectangles is lighter and the right side of each rectangle is darker, despite each rectangle being of uniform color. This is a result of the center-surround receptive fields of ganglion cells, enhancing the contrast between the borders between rectangles.

Q64 What is the Craik-O'Brien-Cornsweet illusion?

A64 In the **Craik-O'Brien-Cornsweet illusion**, the rightmost edge of the image appears darker than the leftmost edge of the image. However, both edges are the same color. Because the brain mainly receives contrast or edge information, the border at the center of the image tricks the brain into believing that the entire right side of the image is darker than the left side.

Wachtler, Thomas, and Christian Wehrhahn. "The Craik—OBrien—Cornsweet Illusion in Colour: Quantitative Characterisation and Comparison with Luminance." *Perception* 26, no. 11 (1997): 1423–430. doi:10.1068/p261423.

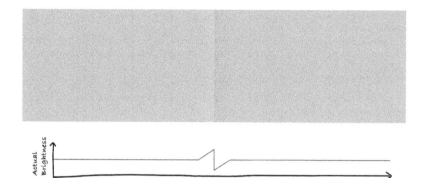

The right rectangle appears overall darker than the left rectangle, despite both rect-angles being the same darkness (except at the border). This can be proven by placing a finger over the border so on the the left and right sides can be seen, showing that the two are in fact the same color. A graph of this effect comparing actual and per-ceived brightness is shown below. This demonstrates that the edge information contrib-utes greatly to the overall perception of an object, and can even result in misleading perception.

Q65 **What are the ON and OFF channels?**

A65 The ON and OFF channels function to detect contrast. Bipolar cells and retinal ganglion cells are divided into ON types and OFF types. The **ON channel** detects light increments, or in other words, where the visual field is brighter than the surrounding. The **OFF channel** detects light decrements, where the visual field is darker than the sur-rounding. For example, the OFF channel is used to read black text on a white background, as the text is darker than the background, while one would rely more on the ON channel to read white text on a black background.

Schiller, Peter H., Julie H. Sandell, and John H. R. Maunsell. "Functions of the ON and OFF channels of the visual system." *Nature* 322, no. 6082 (1986): 824–25. doi:10.1038/322824a0.

Q66 **How do the ON bipolar cells respond to a cone detecting light?**

A66 When a cone detects light, the amount of glutamate released into the synapse decreases. In the synapse with an **ON bipolar cell**, glutamate is an inhibitory neurotransmitter. When the amount of glutamate decreases, bipolar cells are activated. Because of this, the cone detecting light leads to the bipolar cell turning ON, hence the name.

Nelson, Ralph, and Victoria Connaughton. "Bipolar Cell Pathways in the Vertebrate Retina." *University of Utah.* http://webvision.med.utah.edu/book/part-v-phototransduction-in-rods-and-cones/bipolar-cell-pathways-in-the-vertebrate-retina/.

Q67 **How do the OFF bipolar cells respond to a cone detecting light?**

A67 **OFF bipolar cells** react the opposite way compared to ON bipolar cells. In synapses with an OFF bipolar cell, glutamate is excitatory. When the cone detects light, the amount of glutamate released decreases, causing the bipolar cell to turn OFF.

Nelson, Ralph, and Victoria Connaughton. "Bipolar Cell Pathways in the Vertebrate Retina." *University of Utah.* http://webvision.med.utah.edu/book/part-v-phototransduction-in-rods-and-cones/bipolar-cell-pathways-in-the-vertebrate-retina/.

Q68 **How do the ON and OFF channels function with rod cells?**

A68 All bipolar cells receiving input from rods are ON bipolar cells. Rods are not connected to any off bipolar cells. However, they do still contribute to ON and OFF channels. The rod ON bipolar cells synapse with an AII amacrine cell, which in turn synapses both ON and OFF type cells. The synapse with an ON retinal ganglion cell is excitatory, while the synapse with an OFF retinal ganglion cell is inhibitory.

Bach, Michael. "Craik-O'Brien-Cornsweet illusion." *Michael Bach.* June 13, 2002. http://www.michaelbach.de/ot/lum-cobc/index.html.

Q69 **What is an evolutionary advantage of having both ON and OFF channels?**

A69 An advantage of having both ON and OFF channels is advantageous, for it is much more efficient in signalling relatively large variations of light. An example of an evolutionary benefit would be in fish, who have to be able to spot predators from below and above. Predators from below would reflect light, and appear brighter than the background. Predators from above would block light, and appear darker than the background. The ON and OFF channels work together to allow the fish to evade predation.

Chalupa, Leo M., and John S. Werner. *The Visual Neurosciences.* Cambridge, MA: MIT Press, 2003.

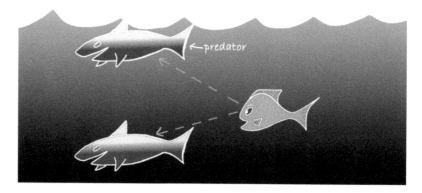

A possible evolutionary advantage of the ON and OFF channels in the visual system is the ability to efficiently perceive regions brighter than the background and regions darker than the background simultaneously. In this example, the fish (orange) is able to see both sharks (turquoise) clearly due to the ON and OFF channels. The shark above is darker than the background, as it blocks sunlight filtering down, while the shark below is brighter than the background, as it reflects sunlight from above.

Q70   What are the types of retinal ganglion cells?

A70   There are four main classes of **retinal ganglion cells**: midget, parasol, bistratified, and intrinsically photosensitive ganglion cells.

Sanes, Joshua R., and Richard H. Masland. "The Types of Retinal Ganglion Cells: Current Status and Implications for Neuronal Classification." *Annu. Rev. Neurosci.* 38, no. 1 (2015): 221–46. doi:10.1146/annurev-neuro-071714-034120.

Q71   What are midget cells?

A71   **Midget cells** are a type of retinal ganglion cell that project to the parvocellular layers of the lateral geniculate nucleus. They are called midget cells due to their small size in cell body and dendrites. These cells compose 80% of all the retinal ganglion cells. They generally have small receptive fields and slow conduction velocities. Midget cells respond in a sustained manner to stimuli; they will fire action potentials for as long as the light is detected.

"Midget cell." *Oxford University Press.* http://oxfordindex.oup.com/view/10.1093/oi/authority.20110803100156522.

Hacking, Craig. "Optic tract." *Radiopaedia.* https://radiopaedia.org/articles/optic-tract.

Dacey, Dennis. "The mosaic of midget ganglion cells in the human retina." *J. Neurosci.* 13, no. 12 (December 1, 1993): 5334–355. https://www.ncbi.nlm.nih.gov/pubmed/8254378.

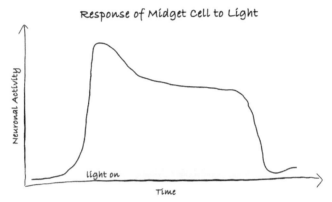

The response of a midget cell to light plots neuronal activity (action potential firing rate) in response to the detection of light. When the light is turned on, the midget cell increases in activity, then plateaus, remaining relatively active until the light turns off.

Q72 What are parasol cells?

A72 Parasol cells are another type of retinal ganglion cells. Unlike midget cells, however, the parasol cells' axons project to the magnocellular layers of the lateral geniculate nucleus. In these layers, the axons of the parasol cells synapse with the magnocellular cells. These cells are called parasol cells due to the shape of the large dendrites. Approximately 10% of all retinal ganglion cells are parasol cells. These cells have large receptive fields and are connected to large amounts of rods and cones. Because parasol cells receive mixed input from red, green, and blue cones, they cannot carry color information. Unlike midget cells, parasol cells have fast conduction velocity, and can also respond to low-contrasts stimuli. Furthermore, parasol cell responses to stimuli are more transient; they fire only when the light turns on or off.

Callaway, Edward M. "Structure and function of parallel pathways in the primate early visual system." *J. Physiol.* 566, no. 1 (2005): 13–19. doi:10.1113/jphysiol.2005.088047.

Watanabe, M., and R. W. Rodieck. "Parasol and midget ganglion cells of the primate retina." *J. Comp. Neurol.* 289, no. 3 (1989): 434–54. doi:10.1002/cne.902890308.

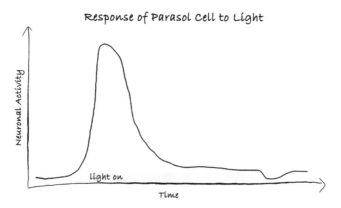

The response of a parasol cell to light plots neuronal activity (action potential firing rate) in response to the detection of light. When the light is turned on, the parasol cell increases in activity, then almost immediately decreases to baseline levels.

Q73 What are bistratified cells?

A73 There are two types of **bistratified cells**: small bistratified ganglion cells, and large bistratified ganglion cells. Bistratified cells compose around 8–10% of retinal ganglion cells and receive information from either bipolar cells or amacrine cells and send them to the koniocellular layers of the LGN. The cells are thought to aid in color vision. Some distinguishable characteristics of the cells include: moderate conduction velocity, moderate spatial resolution, and also response to moderate contrast stimuli.

Daw, Nigel. *How Vision Works.* New York: Oxford University Press, 2012.

Dacey, Dennis M. "Morphology of a small-field bistratified ganglion cell type in the macaque and human retina." *Vis. Neurosci.* 10, no. 06 (1993): 1081–098. doi:10.1017/s0952523800010191.

Q74 What are intrinsically photosensitive ganglion cells?

A74 Intrinsically photosensitive ganglion cells are a type of neuron in the retina. These types of ganglion cells are special, for they are the only type of retinal cell, besides rods and cones, that are photosensitive. These cells are a third type of retinal photoreceptors.

Do, M. T. H., and K.-W. Yau. "Intrinsically Photosensitive Retinal Ganglion Cells." *Physiol. Rev.* 90, no. 4 (2010): 1547–581. doi:10.1152/physrev.00013.2010.

Q75   **What is the midget system?**

A75   The **midget system** is a pathway stemming from the midget ganglion cells of the retina that extends to the various areas of the cerebral cortex. The midget cells receive inputs from a small number of photoreceptors, leading to higher spatial acuity. These cells synapse with parvocellular cells of the lateral geniculate nucleus, which then convey the information to the visual cortex for processing.

The midget system processes high spatial frequency information (more detail) and color. Monkeys with lesions in the parvocellular layers of the LGN are unable to discriminate between colors and show decreased visual acuity.

However, because midget ganglion cells respond in a sustained fashion to stimulus, they are unsuitable for processing of motion information.

Kolb, Helga. "Midget pathways of the primate retina underlie resolution and red green color opponency." *University of Utah*. June 2012. http://webvision. med.utah.edu/book/part-iii-retinal-circuits/midget-pathways-of-the-primate-retina-underly-resolution/.

Q76   **What is the parasol system?**

A76   The parasol system is another pathway originating from the parasol ganglion cells of the retina. These ganglion cells have large receptive fields, leading to lower spatial acuity. However, they respond transiently, firing for a short period of time in response to a change in color or brightness. Parasol cells synapse with magnocellular cells of the lateral geniculate nucleus, which convey the information to the visual cortex for processing.

The parasol system processes high temporal frequency information, such as motion. Monkeys with lesions in the magnocellular layers of the LGN show an inability to perceive motion and quick changes in color or brightness, such as flickers.

However, due to the large receptive fields of parasol cells, the parasol system is unable to provide detailed or color information.

Schiller, Peter H., and Edward J. Tehovnik. "The Midget and Parasol Systems." *Vision and the Visual System*, 2015, 135–58. doi:10.1093/acprof: oso/9780199936533.003.0008.

4 Subcortical Structures

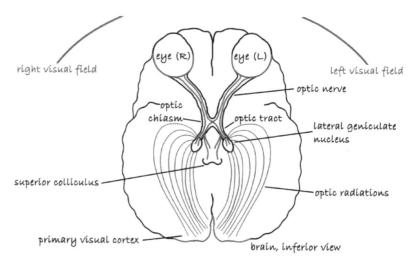

The visual pathway from the visual field to the visual cortex is shown, with red representing the right half of the visual field and blue representing the left half of the visual field. The image is inverted when projected onto the retina, then sent via the optic nerves to the brain. At the optic chiasm, some nerve fibers cross so that information from one side of the visual field is sent to the opposite side of the brain. The nerve fibers then form the optic tract and synapse at the lateral geniculate nucleus, as well as other midbrain structures. From there, optic radiations carry the visual information to the primary visual cortex.

Q77 What is the optic chiasm?

A77 The two optic nerves, one from each eye, intersect at the **optic chiasm**. Some nerve fibers cross over the midline of the brain, and others do not. Past the optic chiasm, the left side of the brain receives visual information from the left side of each eye, while the right side of the brain receives visual information from the right side of each eye.

"Optic Nerves And Chiasm." *Ophthalmology Training*. http://www.ophthalmology-training.com/ocular-anatomy/visual-pathway/optic-nerves-and-chiasm.

Larsson, Matz. "The optic chiasm: a turning point in the evolution of eye/hand coordination." *Front. Zool.* 10, no. 1 (2013): 41. doi:10.1186/1742-9994-10-41.

Q78 What are the optic tracts?

A78 The **optic tract** is a continuation of the optic nerve from the optic chiasm to the lateral geniculate nucleus, pretectal nuclei, and superior colliculus.

"Optic Nerve." *California State University, Northridge.* http://www.csun. edu/~vcpsy00i/dissfa01/optic.html.

Q79 **How does a lesion of a single optic tract (left or right) affect a person's ability to see?**

A79 Lesions in the optic tract lead to visual field loss on either the right or left of the vertical midline. This is called **homonymous hemianopsia**. A lesion on the right optic tract will lead to loss of vision on the left half of the visual field, while a lesion on the left optic tract will lead to loss of vision on the right side. This condition usually affects both eyes, but can affect just one eye. In most cases of hemianopsia, there is central sparing, where central vision remains unaffected. Anopsia is a general term for a defect in the visual field.

"Homonymous Hemianopsia." *Cleveland Clinic.* April 17, 2015. https:// my.clevelandclinic.org/health/articles/homonymous-hemianopsia.

"What is hemianopia?" *University College London.* http://www.eyesearch.ucl.ac.uk/ es/es_hemianopia.php.

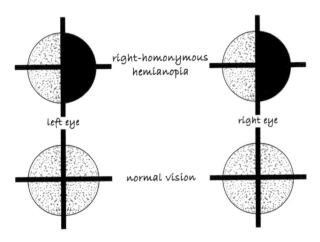

In right homonymous hemianopia (top), no information is sent to the brain from the right half of the visual field, from either eye. As a result, the patient afflicted is blind on the right side. Note that in most cases, central vision is unaffected.

Q80 **Where do the retinal ganglion cells synapse in the brain?**

A80 Retinal ganglion cells synapse at the thalamus, hypothalamus, and superior colliculus in the brain.

"Retinal Ganglion Cell." *NeuronBank*. September 13, 2009. http://neuronbank.org/wiki/index.php/Retinal_Ganglion_Cell.

Q81 **What is the lateral geniculate nucleus?**

A81 The **lateral geniculate nucleus**, or LGN, is a paired nucleus within the thalamus that relays all visual information from the eyes to the visual cortex for further processing. There are two LGNs, one on each of the brain. The LGN can be separated into three groups: the magnocellular layers, parvocellular layers, and koniocellular neurons.

Dragoi, Valentin. "Chapter 15: Visual Processing: Cortical Pathways." *Neuroscience Online*. Accessed September 9, 2017. http://neuroscience.uth.tmc.edu/s2/chapter15.html.

Q82 **How is the lateral geniculate nucleus (LGN) structured?**

A82 The lateral geniculate nucleus is structured as six principal layers of neurons, one through six, where layer one is the innermost layer and layer six is the outermost layer. The large magnocellular cells form the deepest two layers of the LGN, layers 1 and 2, while the upper four layers, 3 through 6, are composed of smaller parvocellular cells. Thin layers of the smallest cells, koniocellular neurons, are interposed between the six principal layers. The axons of these cells end in different layers and sublayers of the primary visual cortex. Furthermore, layers 2, 3, and 5 receive input from the ipsilateral eye, which is the eye on the same side of the body. Layers 1, 4, and 6 receive visual information from the contralateral eye, or the eye on the other side of the body.

Mather, George. "The Lateral Geniculate Nucleus." *University of Sussex*. Accessed September 9, 2017. http://www.lifesci.sussex.ac.uk/home/George_Mather/Linked%20Pages/Physiol/LGN.html.

Q83 **What are parvocellular cells?**

A83 **Parvocellular cells** receive visual input from midget ganglion cells of the retina, and thus are part of the midget system. As a result, the midget system is also called the parvo or parvocellular system.

Dragoi, Valentin. "Chapter 15: Visual Processing: Cortical Pathways." *Neuroscience Online*. Accessed September 9, 2017. http://neuroscience.uth.tmc.edu/s2/chapter15.html.

Q84 **What are magnocellular cells?**

A84 **Magnocellular cells** receive input from parasol ganglion cells of the retina, and thus are part of the parasol system. As a result, the parasol system is also called the magno or magnocellular system.

Dragoi, Valentin. "Chapter 15: Visual Processing: Cortical Pathways." *Neuroscience Online.* Accessed September 9, 2017. http://neuroscience.uth.tmc.edu/s2/chapter15. html.

Q85 **What are koniocellular cells?**

A85 **Koniocellular cells** are another type of neuron in the lateral geniculate nucleus. The specific function of koniocellular cells is unknown, but there are various hypotheses regarding their function. They are thought to synapse in the blobs of the primary visual cortex.

Dragoi, Valentin. "Chapter 15: Visual Processing: Cortical Pathways." *Neuroscience Online.* Accessed September 9, 2017. http://neuroscience.uth.tmc.edu/s2/chapter15. html.

Q86 **What are the optic radiations?**

A86 The **optic radiations** send information from the LGN to the visual cortex. Optic radiations in each hemisphere are divided into the lower and upper division, known as **Meyer's loop** and **Baum's loop**, respectively. The two Meyer's loops (left and right) carry information from the lower half of the retina, while the two Baum's loops carry information from the upper half. Thus, information from the upper right quarter of the visual field is transferred through the left Meyer's loop, and information from the lower left quarter of the visual field is transferred through the right Baum's loop.

Lisik, James. "Optic radiation." *Radiopaedia.* Accessed September 14, 2017. https:// radiopaedia.org/articles/optic-radiation-1.

Chowdhury, F. H., and A. H. Khan. "Anterior & lateral extension of optic radiation & safety of amygdalohippocampectomy through middle temporal gyrus: a cadaveric study of 11 cerebral hemispheres." *Asian J Neurosurg.* 5, no. 1 (January 2010): 78–82. Accessed September 21, 2017. https://www.ncbi.nlm.nih.gov/pubmed/22028747.

Q87 **How does a lesion of a single optic radiation affect a person's ability to see?**

A87 If the lesion only occurs on a single optic radiation, processing will not occur for the single quadrant of vision that the axon covers, which is known as **quadrantanopia**. So, vision will be lost for one of the 4 quadrants that in total define a human's total vision. The affected quadrant will seem pale and foggy, indistinct compared to other areas. To figure out which optic radiation was affected, it is important to understand the mixing of optic radiations. Because of the switching of left visual field to the right side of the brain, and because of the different pathways of lower half and upper half images, the quadrant that vision is affected is imperative in figuring out where the lesion has occurred. For example, vision impairment of the upper left quadrant of the visual field, therefore being the lower right quadrant of the retinas, would mean that the lesion would have occurred in the right Meyer's loop. Or, vision impairment of the lower left quadrant of the visual field, therefore being the upper right quadrant of the retinas, would mean that a lesion has occurred in the right Baum's loop. The opposites of these cases are also true.

Purves, D., G. J. Augustine, and D. Fitzpatrick. "Visual Field Deficits." *Neuroscience.* 2nd ed. Sunderland, MA: Sinauer, 2001. Accessed September 20, 2017. https://www.ncbi.nlm.nih.gov/books/NBK10912/.

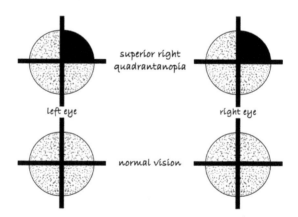

In superior right quadrantanopia (top), no information is sent to the brain from the upper right quadrant of the visual field, from either eye. As a result, the patient afflicted is blind in the top-right quarter of the visual field.

Q88 **Where else in the brain, besides the LGN, do the fibers of the optic nerve synapse?**

A88 Besides the LGN, the retinal ganglion cells synapse at various other locations, such as the superior colliculi, the pretectal area, and the suprachiasmatic nucleus.

Dragoi, Valentin. "Chapter 15: Visual Processing: Cortical Pathways." *Neuroscience Online.* Accessed September 9, 2017. http://neuroscience.uth.tmc.edu/s2/chapter15.html.

Q89 **What is the superior colliculus?**

A89 The **superior colliculus** is a structure involved in visual reflexes that is located in the tectum of the midbrain. In mammals, it is split into 7 layers or lamina. The outer 3 layers are the superficial lamina which receive input from visual areas. The middle 2 layers and the inner 2 layers make up the intermediate and deep lamina, respectively. These 4 layers receive input from various sensory and motor areas that are useful in a reflex action. The major function of the superior colliculus in vision is to allow the eyes to fixate, saccade and converge. Fixation is the movement of the eye towards an object, without moving the head. Saccades are movements where the eye moves rapidly, changing focus multiple times. Vergence describes the movement of each eye in opposite directions to maintain focus.

"Superior Colliculus." *Neuroscience Resource Page.* Accessed September 21, 2017. http://www.neuroanatomy.wisc.edu/Bs97/TEXT/P23/ov.htm.

Ghandi, Neeraj J., and Husam A. Katnani. "Motor Functions of the Superior Colliculus." *Annu. Rev. Neurosci.* 34 (July 2011): 205–31. Accessed September 21, 2017. doi:10.1146/annurev-neuro-061010-113728.

Q90 **What is the pretectal area?**

A90 The **pretectal area**, also known as the pretectal nuclei, is a region composed of multiple nuclei located in front of the superior colliculi. This area receives information from the retinal ganglion cells via the optic tract, and is involved in the pupillary light reflex.

"Pretectal nuclei." *MediLexicon.* Accessed September 14, 2017. http://www.medilexicon.com/dictionary/61672.

Q91 **What is the pupillary light reflex?**

A91 The **pupillary light reflex** is where the size of the pupil changes depending on the amount of light available. For example, high levels of light decreases the size of the pupil to limit entry of excess amounts of light. On the other hand, when there is low luminance, the pupils dilate to let in as much light as possible.

"Pupillary Light Reflex Pathway." *Medical Institution.* Accessed September 14, 2017. http://www.medical-institution.com/pupillary-light-reflex-pathway/.

Q92 **What is the suprachiasmatic nucleus?**

A92 The **suprachiasmatic nucleus** (SCN) is a part of the **hypothalamus**, a structure of the brain located below the thalamus. One of the functions of the hypothalamus is maintaining homeostasis, or a constant internal environment. The SCN controls the circadian rhythm.

"The Human Suprachiasmatic Nucleus." *HHMI BioInteractive*. Accessed September 14, 2017. http://www.hhmi.org/biointeractive/human-suprachiasmatic-nucleus.

Q93 **What role does vision play in the circadian rhythm?**

A93 In addition to the three types of retinal ganglion cells that were discussed, there is another type called intrinsically photosensitive retinal ganglion cells (ipRGC). These cells are similar to photoreceptors in that they contain photopigments, in this case **melanopsin**, which allows them to detect light. The ipRGCs then send axons to the suprachiasmatic nucleus (SCN) in the hypothalamus, which is responsible for regulating the **circadian rhythm**, which is the daily 24-hour cycle pertaining to the physiological processes of an organism. The SCN uses this input to match the circadian rhythm to the cycle of day and night.

"The Human Suprachiasmatic Nucleus." *HHMI BioInteractive*. Accessed September 14, 2017. http://www.hhmi.org/biointeractive/human-suprachiasmatic-nucleus.

Q94 **What is blindsight?**

A94 **Blindsight** is a condition in which someone is able to respond to visual stimuli without being able to consciously see those stimuli. There are two ways in which this occurs. Type 1 blindsight involves the ability of those cortically blind in only one side of the visual field to detect and differentiate between stimuli on their blind side much more accurately than one could simply through guessing. Type 2 blindsight is when someone feels something change in their blind area that has not been visually detected. They may not know what has changed, but they can sense that something has changed without conscious visual perception of the region.

Robson, David. "Blindsight: the strangest form of consciousness." *BBC*. September 28, 2015. Accessed September 14, 2017. http://www.bbc.com/future/story/20150925-blindsight-the-strangest-form-of-consciousness.

Q95 **What is the Anton-Babinski syndrome?**

A95 **Anton-Babinski syndrome,** also known as visual anosognosia, is a rare symptom where damage to the occipital lobe results in the person who is clearly blind but still believes that he or she can still see. The patients will often make up details about their surroundings and refuse to admit that they are blind. The reason those afflicted with the syndrome deny their blindness is currently not known but there are numerous hypotheses. A possible hypothesis is that damage to the visual cortex results in an inability to communicate with the speech portion of the brain thus the speech region makes up a response.

Chen, Jiann-Jy, Hsin-Feng Chang, Yung-Chu Hsu, and Dem-Lion Chen. "Anton-Babinski syndrome in an old patient: a case report and literature review." *Psychogeriatrics* 15, no. 1 (March 2015): 58–61. Accessed September 21, 2017. doi:10.1111/psyg.12064.

5 Cortical Circuitry

Q96 **What is the great-grandmother cell hypothesis?**

A96 The **great-grandmother cell hypothesis** is a name for an early attempt to explain how the brain processes visual information. In this hypothesis, there is a neuron in the brain for every object that can be recognized, including one for your great-grandmother. However, this hypothesis has been refuted, as a neuron for every recognizable object means that the brain would have to be far larger than is realistic.

Gross, C. G. "Genealogy of the 'grandmother cell.'" *Neuroscientist* 8, no. 5 (October 2002): 512–18. Accessed September 21, 2017. https://www.ncbi.nlm.nih.gov/pubmed/12374433.

Q97 **What is the cerebral cortex?**

A97 The **cerebral cortex** (or "cortex") is a layer of neurons two to four millimeters thick on the surface of the brain. It is composed closely packed neurons forming gray matter. The cerebral cortex consists of four lobes: the occipital, parietal, temporal, and frontal lobes. It is divided into the left and right cerebral hemispheres.

Some main functions of the cerebral cortex include perception, awareness, memory, consciousness, speech, and higher level thinking.

"The Cerebral Cortex." *AP Psychology Community*. Accessed September 14, 2017. http://www.appsychology.com/Book/Biological/cerebral_cortex.htm.

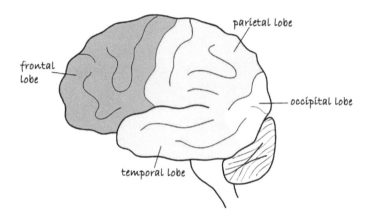

The cerebral cortex consists of four lobes: the frontal lobe (red), parietal lobe (green), temporal lobe (blue), and occipital lobe (yellow).

Q98 **What is the function of each lobe of the cortex?**

A98 Each lobe in the cortex serves unique roles. The **frontal lobe** is considered to be the control center for emotions and personality. It is also involved in problem solving, motor skills, judgement, and social behaviour. The **parietal lobe** processes information related to several senses, including taste, temperature, touch, as well as helping out with vision. The **temporal lobe** is primarily involved in hearing and language. However, parts of the temporal lobe are also dedicated to vision. Finally, the **occipital lobe** primarily processes vision.

Vision is an important sense in humans, as the entire occipital lobe as well as parts of the parietal and temporal lobes are dedicated to processing vision.

"Frontal Lobes." *Centre for Neuro Skills*. Accessed September 14, 2017. http://www.neuroskills.com/brain-injury/frontal-lobes.php.

"Temporal Lobes." *Centre for Neuro Skills*. Accessed September 14, 2017. http://www.neuroskills.com/brain-injury/temporal-lobes.php.

"Parietal Lobes." *Centre for Neuro Skills*. Accessed September 14, 2017. http://www.neuroskills.com/brain-injury/parietal-lobes.php.

"Occipital Lobes." *Centre for Neuro Skills*. Accessed September 14, 2017. http://www. neuroskills.com/brain-injury/occipital-lobes.php.

Q99 **Where in the cerebral cortex do the optic radiations synapse?**

A99 The optic radiations synapse in the primary visual cortex, which is also called V1, the striate cortex, or Brodmann's area 17.

Loh, Daniel. "Cuneus." *Radiopaedia*. Accessed September 14, 2017. https://radio paedia.org/articles/cuneus.

Loh, Daniel. "Lingual gyrus." *Radiopaedia*. Accessed September 14, 2017. https:// radiopaedia.org/articles/lingual-gyrus.

Q100 **What is the primary visual cortex?**

A100 The **primary visual cortex** is the location of the first stage in process-ing visual information. It contains a complete map of the visual field from the lateral geniculate nucleus. V1 is also thought to filter out the visual information to enhance the contours of the image, which allows the brain to distinguish different objects.

Carandini, Matteo. "Area V1." *Scholarpedia* 7, no. 7 (July 2012). Accessed September 21, 2017. doi:10.4249/scholarpedia.12105.

Q101 **How is the primary visual cortex structured?**

A101 The primary visual cortex (V1) is divided into 6 horizontal layers, dubbed 1–6. Layer 1 is the outermost layer, while layer 6 is the deep-est layer. The thickest nerve bundles from the LGN project to layers 5 and 6, which are the layers of the V1 closest to the center of the brain. Though layers 5 and 6 receive much input, it is actually layer 4 of V1 that receives the most visual input from the LGN, so much that it actually has 4 sublayers, 4A, 4B, 4Cα, and 4Cβ. In terms of input, sublamina 4Cα receives widely magnocellular input, while sublamina 4Cβ receives mostly parvocellular input which then goes to blobs for processing.

Schmolesky, Matthew. "The Primary Visual Cortex." *In Webvision: The Organization of the Retina and Visual System.*, edited by H. Kolb, E. Fernandez, and R. Nelson. Salt Lake City, UT: University of Utah Health Sciences Center, n.d. Accessed September 21, 2017. https://www.ncbi.nlm.nih.gov/books/NBK11524/.

Q102 **What is a retinotopic map?**

A102 A **retinotopic map** is a method of organization of neurons in the visual system. In the visual cortex, as well as other areas of the visual pathway, neurons are arranged so that their receptive fields form a map that reflects the spatial organization of the visual field. For example, if a person were presented with a square in the center of the visual field, the corresponding square-shaped region of the visual cortex would fire in response. A triangle in the visual field would lead to a triangular region firing in the visual cortex, and so on. However, this does not mean that the organization of the visual cortex is exactly the same as the visual field, since there is significant distortion of the retinotopic map. Firstly, because images are inverted vertically and horizontally as they pass through the cornea and lens, the image that forms on the retina is upside-down, as is the visual cortex. Due to the organization of the optic chiasm and the optic tracts, the left half of the visual cortex deals with the right visual field, and the right visual cortex deals with the left visual field. Furthermore, because the fovea contains the highest visual acuity, there are many more neurons dedicated to the processing of central vision, leading to more representation of the foveal input in the visual cortex and less representation of peripheral vision.

Saha, Sanna. "Retinotopic Maps." *University of Southampton*. Accessed September 14, 2017. http://users.ecs.soton.ac.uk/harnad/Hypermail/Explaining.Mind/0167.html.

Wang, Ruye. "Retinotopic Mapping." *Harvey Mudd College*. August 4, 2013. Accessed September 14, 2017. http://fourier.eng.hmc.edu/e180/lectures/v1/node3.html.

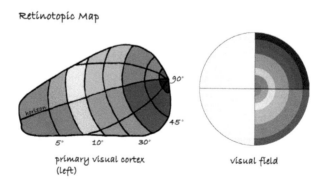

Retinotopic Map

primary visual cortex (left)

visual field

The retinotopic map is how each point on the visual field is mapped in terms of neurons in the visual pathway. Although general spatial relationships are preserved, the retinotopic map also introduces heavy distortion. For example, due to its high visual acuity, there are far more neurons dedicated to central vision than there are for peripheral vision.

Q103 **What are the types of cells in the primary visual cortex?**

A103 In the visual cortex, there are three basic types of cells: simple, complex, and end-stopped cells.

Heeger, David. "Perception Lecture Notes: LGN and V1." *Center for Neural Science.* Accessed September 21, 2017. http://www.cns.nyu.edu/~david/courses/perception/lecturenotes/V1/lgn-V1.html.

Q104 **What are simple cells?**

A104 **Simple cells** have receptive fields similar to the center-surround receptive fields in the retina and LGN. There is a region where stimulus excites the cell, which is surrounded by regions where stimulus inhibits the cell. However, in a simple cell, the excitatory region is rectangular rather than circular. Furthermore, simple cell receptive fields are also sensitive to orientation. When a bar of light is presented within the receptive field of a simple cell, the cell only responds if the bar of light is at the right orientation.

Hubel, David H. "Simple Cells." *Scientific American Library*, 1988. Accessed September 21, 2017. http://hubel.med.harvard.edu/book/b17.htm#simp.

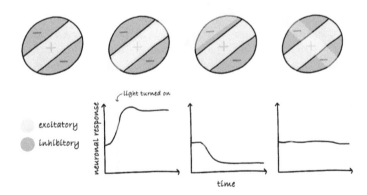

A simple cell has an excitatory region surrounded by inhibitory regions. When light is shined on the excitatory region, neuronal activity increases. When light is shined on the inhibitory region, neuronal activity decreases. Simple cells often also exhibit orientation sensitivity, as if a bar of light is not at the correct orientation, the neuron does not respond.

Q105 **What are complex cells?**

A105 Unlike simple cells, **complex cells** do not respond to a static bar of light, but rather a bar of light moving across its receptive field.

Complex cells are sensitive to the orientation of the bar of light and the direction of its motion. Furthermore, complex cells do not have designated excitatory and inhibitory regions. Complex cells respond to the stimulus as long as it is oriented at the correct angle and traveling in the correct direction, regardless of its actual location within the receptive field.

Hubel, David H. *Eye, Brain, and Vision*. Scientific American Library, 1988. Accessed September 21, 2017. http://hubel.med.harvard.edu/.

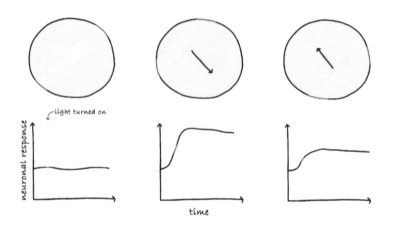

Complex cells are sensitive to orientation as well as motion. A stationary bar of light elicits no response, but if the bar of light is moving, the neuron responds. Complex cells are also sensitive to the direction of motion, as the neuron does not respond as strongly if the bar of light is moving in the opposite direction.

Q106 **What are end-stopped cells?**

A106 **End-stopped cells**, also known as **hypercomplex cells**, demonstrate a property known as end-stopping, or **end-inhibition**, in which the response of the cell decreases as the size of the stimulus increases. For example, when an end-stopped cell is presented with a bar of light at the right orientation, it responds, similar to a simple cell or complex cell. However, when the length of the bar of light increases, the response of the end-stopped cell decreases.

It was originally thought that end-stopped cells were a class of cells distinct from simple and complex cell, and that simple and complex cells do not demonstrate end-stopped behavior. However, it has since been discovered that some simple cells and some complex cells do show end-stopping. Thus, end-stopped cells are now classified as a subtype of simple and complex cells, i.e. simple end-stopped cells and complex end-stopped cells.

Hubel, David H. *Eye, Brain, and Vision*. Scientific American Library, 1988. Accessed September 21, 2017. http://hubel.med.harvard.edu/.

Q107   **What is the motion after effect illusion, and how does it work?**

A107   An example of the **motion aftereffect illusion** can be found here: https://strobe.cool/. In this illusion, the viewer fixates at a spot on the screen while there is constant motion in a single direction. After a period of time, the user looks away from the screen and sees distortions near the center of the field of vision, which gradually fades away over a period of about a minute.

This illusion works by adapting the complex cells of the visual cortex. Consider cell A, a complex cell tuned in the same direction as the motion of the bars of light on the screen, while cell B is tuned in the opposite direction. Normally, both cells A and B have a low baseline activity, which balance out. When presented with the illusion, cell A increases in activity. After viewing the illusion for an extended period of time, cell A adapts to the constant motion, and thus becomes desensitized, firing less in response to motion. Cell B, on the other hand, continues its low baseline firing rate without change when presented with the illusion. When the user looks away from the illusion, cell B is still fires at its low baseline amount. However, cell A now is desensitized and fires at a lower rate than baseline. The firing rates of cells A and B no longer balance each other out, creating a sense of motion where there actually is none.

"Motion Aftereffect." *Illusion Works*. Accessed September 21, 2017. https://psy2.psych.tu-dresden.de/i1/kaw/diverses%20Material/www.illusionworks.com/html/motion_aftereffect.html.

Q108   **What are "blobs"?**

A108   In 1979, Margaret Wong-Riley stained the visual cortex with an enzyme called **cytochrome oxidase**. Certain regions became stained and other regions did not stain. The regions that stain from cytochrome oxidase were named "**blobs**". These cells in the blobs are thought to be involved in color processing, as over half of all blob cells are color responsive.

Hubel, David H. "Blobs and Color Vision." In *Bioscience at the Physical Science Frontier*, edited by Claudio Nicolini, 91–102. Humana Press, 1986. Accessed September 21, 2017. doi:10.1007/978-1-4612-4834-7_6.

Q109   **What is the Hering opponent process theory?**

A109   The trichromatic theory states that color is processed using three primary colors, red, green, and blue. This theory explains how cone photoreceptors contribute to the ability to distinguish color but does not explain certain phenomena related to color perception. Some theorists argued that color is processed at the ganglion cell level, one of which was Ewald Hering who proposed the **opponent process theory**. The theory states that cone photoreceptors are linked together by to form pairs of opposite colors. These pairs are blue/ yellow, red/green, and black/white. Activation of member inhibits the activation of the other. Hering argued that because the colors "bluish-yellow" and "reddish-green" do not exist, blue/yellow and red/green must be opposite and mutually exclusive. The two strongest pieces of evidence to support this theory can be found in color blindness and afterimages.

Kline, Don. "Theories of Color Vision." *Vision & Aging Lab.* 1997. Accessed September 14, 2017. http://psych.ucalgary.ca/PACE/VA-Lab/colourperceptionweb/ theories.htm.

Q110   **What are afterimages?**

A110   When the eyes are constantly fixated at a specific object, slowly the retinal cells become fatigued and desensitized to light. The desensitization is strongest for photoreceptors viewing the brightest part of an image and least for those that view the darkest part. After staring at an image, suddenly changing what you're looking at will cause the creation of an afterimage because the less fatigued cells experience light as more bright than the more desensitized cells even if there is no real change in brightness.

For example, let's say you have a green square, and you stare at for a whole minute. This causes the green-sensitive retinal cells to become fatigued. As a result, when you shift focus to any blank surface, the green cells will be temporarily unable to fire. The opponent cells then receive a decreased amount of green input, and since red and green are opposing colors, this increases the perception of red, creating a red afterimage.

"Afterimage." *IllusionWorks*. Accessed September 21, 2017. http://psylux.psych. tu-dresden.de/i1/kaw/diverses%20Material/www.illusionworks.com/html/ afterimage.html.

Kendra Cherry. "What Is the Opponent Process Theory of Color Vision?" *Verywell*. Accessed September 14, 2017. https://www.verywell.com/what-is-the-opponent-process-theory-of-color-vision-2795830.

Stare at the black dot at the center of the square for about 30 seconds, then view the white region to the right of it. You should see a reddish afterimage. The color of the afterimage is explained by the opponent theory. Staring at the green square fatigues the neurons responsible for transmitting green signal, causing them to be less active. When the white region is viewed, the red and green signals that normally balance out in the opponent process are no longer in balance, since the green input is decreased. As a result, a red afterimage is perceived.

Q111 **Which theory is correct: the trichromatic theory or the opponent theory?**

A111 The trichromatic theory and the opponent theory are both correct. The trichromatic theory explains the relevance of cone photoreceptors in processing color while the opponent theory explains the processing of color in the visual cortex.

Kline, Don. "Theories of Color Vision." *Vision & Aging Lab.* 1997. Accessed September 14, 2017. http://psych.ucalgary.ca/PACE/VA-Lab/colourperceptionweb/theories.htm.

Q112 What are single-opponent cells?

A112 Single-opponent cells are a group of neurons located in the blobs of V1 that are characterized by receiving two different colored inputs. There are two major types of single-opponent cells based on the wavelengths of the light they receive: red-green and blue-yellow.

Red-green cells receive input from both L (red) and M (green) cones. Blue-yellow opponent cells receive input from both S (blue), M (green), and L (red) cones. In these cells, the red and green inputs are summed, representing yellow. The sum of the two cone inputs are then opposed to the blue input, forming blue-yellow antagonism. These single-opponent cells create the basis of Hering's opponent theory.

Shapley, Robert, and Michael J. Hawken. "Color in the Cortex: single- and double-opponent cells." *Vision Res.* 51, no. 7 (April 2011): 701–17. Accessed September 21, 2017. doi:10.1016/j.visres.2011.02.012.

Q113 Since there are very few blue cone photoreceptors in the fovea, how is the color blue able to be perceived?

A113 Only about 2% of the cones in the retina are S (blue) cones, and almost none of the cones in the fovea are blue-sensitive, so it would be natural to conclude that the eye is far less sensitive to the color blue than to other colors. However, it is obvious that the eye can detect blue just as well as other colors, even in central vision, where there are very few blue cones. This is explained by the opponent process. Based on the opponent process, blue and yellow are antagonistic, so very little yellow input to a blue-yellow opponent cell leads to an amplified perception of blue color.

"Rods and Cones." *HyperPhysics.* Accessed September 21, 2017. http://hyperphysics.phy-astr.gsu.edu/hbase/vision/rodcone.html.

Q114 What are double-opponent cells?

A114 Double-opponent cells is a particular group of opponent cells. Double-opponent cells have receptive fields where both center and surround regions receive input from single-opponent cells. These cells help to detect colored edges to distinguish different objects from each other, which helps with the perception of the color of an object.

Schwartz, Bennet L., and John H. Krantz. *Sensation and Perception*. Los Angeles: Sage, 2016.

Q115 **What is color constancy?**

A115 **Color constancy** is the ability of humans to perceive colors of objects regardless of the color of the light source. As the surrounding light changes, the color of the light coming from the object changes. However color constancy is the property of the brain in which it compensates for the lighting condition for an accurate perception of the color of the object. For example, imagine you have a red strawberry. Under green light, the strawberry would theoretically have a greenish tint. However, double-opponent cells are thought to take into account the fact that the surroundings are green, causing the strawberry to appear the same red under white and green lights.

"Color Constancy." *AlleyDog*. Accessed September 14, 2017. https://www.alleydog.com/search-results.php?q=color+constancy&x=0&y=0.

"Colour Constancy." *University of East Anglia*. Accessed September 21, 2017. http://www2.cmp.uea.ac.uk/Research/compvis/ColourConstancy/ColourConstancy.htm.

 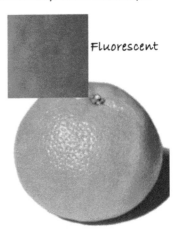

Under incandescent lighting (left) and under fluorescent lighting (right), the oranges are slightly different shades, with the left one appearing brighter and more saturated. These images were taken with a camera using fixed white balance. In real life, the orange under the two lighting conditions would appear nearly identical due to the double-opponent cells and color constancy.

Q116 **What are "interblobs"?**

A116 Unlike blobs, the **interblobs** are regions in the primary visual cortex that do not stain from cytochrome oxidase. These areas are thought to be involved in processing the form and shape of an object. Many neurons of the interblobs demonstrate orientation selectivity, which contributes to form perception.

Wang, Ruye. "Parallel pathways in V1." *Harvey Mudd College*. Accessed September 21, 2017. http://fourier.eng.hmc.edu/e180/lectures/v1/node9.html.

Q117 **Who are David Hubel and Torsten Wiesel?**

A117 David Hubel and Torsten Wiesel were pioneering neurophysiologists who won the Nobel prize in 1981 for Medicine for discovering the functional organization and physiology of neurons that are located in V1.

Heeger, David. "Perception Lecture Notes: LGN and V1." Center for Neural Science. Accessed September 21, 2017. http://www.cns.nyu.edu/~david/courses/perception/lecturenotes/V1/lgn-V1.html.

Nobel Media AB. "David H. Hubel - Facts." *Nobelprize.org*. Accessed September 21, 2017. https://www.nobelprize.org/nobel_prizes/medicine/laureates/1981/hubel-facts.html.

Nobel Media AB. "Torsten N. Wiesel—Facts." *Nobelprize.org*. Accessed September 21, 2017. https://www.nobelprize.org/nobel_prizes/medicine/laureates/1981/wiesel-facts.html.

Q118 **What are ocular dominance columns?**

A118 Hubel and Wiesel discovered that in the primary visual cortex, there are alternating bands of cells which receive input from a certain eye. These bands are known as **ocular dominance columns**. For example, one band running through the visual cortex would receive visual input from the right eye, and surrounding this band would be two other bands receiving input from the left eye.

Furthermore, it was later discovered that blobs are located at the center of ocular dominance columns but never on the border between two ocular dominance columns.

Hubel, David H. *Eye, Brain, and Vision*. Scientific American Library, 1988. Accessed September 21, 2017. http://hubel.med.harvard.edu/.

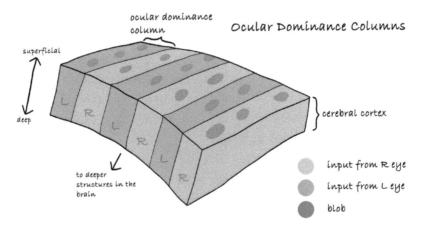

Ocular dominance columns are bands that run along the surface of the visual cortex, where neurons of a single ocular dominance column receive input from a single eye. The ocular dominance columns alternate between left and right eyes. Blobs from cytochrome oxidase staining have been noticed to appear at the center of ocular dominance columns, never between two columns.

Q119   What are orientation columns?

A119   Hubel and Wiesel also discovered that in addition to ocular dominance columns there are also organized regions in the primary visual cortex where the receptive fields of all the cells are selective to the same orientation. These areas are called **orientation columns**. Although the term orientation "columns" suggests a cylindrical shape, this is and true for orientational columns are flat slabs that are perpendicular to the surface of the V1.

Hubel, David H. *Eye, Brain, and Vision.* Scientific American Library, 1988. Accessed September 21, 2017. http://hubel.med.harvard.edu/.

Q120   How are blobs, ocular dominance columns, and orientation columns spatially arranged relative to each other?

A120   Hubel and Wiesel originally proposed what is known as the **"ice-cube" model**, where ocular dominance columns and orientation columns run perpendicularly, creating a grid in the primary visual cortex. Since then, however, the ice-cube model has been shown to be an inaccurate representation of the geometry of the ocular dominance and orientation columns. Instead, a **"pinwheel" model**

was proposed. In this model, each ocular dominance column is divided into segments, each centered on a blob. The orientation columns are then arranged radially (like a pinwheel) around the blob. With new technologies, the geometry of the columns was determined empirically. The orientation columns were discovered to be arranged in a pinwheel-like pattern (but much less organized), often but not always centered on blobs.

Hubel, David H. *Eye, Brain, and Vision.* Scientific American Library, 1988. Accessed September 21, 2017. http://hubel.med.harvard.edu/.

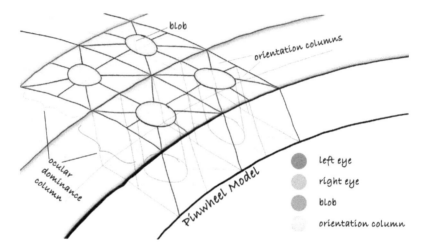

In the pinwheel model, orientation columns (shown in green) radiate outward around each blob, which contain mostly orientation-insensitive neurons. Each orientation column contains neurons sensitive to the same orientation.

Q121   **What are higher-order visual areas?**

A121   **Higher-order visual areas** refer to the extrastriate cortical areas, where processing of color, shape, and motion takes place. Extrastriate cortical areas are regions in the cortex which are located outside of V1 (the striate cortex). Areas such as V2, V3, V4, and V5 are examples of main extrastriate cortical areas.

Dragoi, Valentin. "Chapter 15: Visual Processing: Cortical Pathways." *Neuroscience Online.* Accessed September 9, 2017. http://neuroscience.uth.tmc.edu/s2/chapter15.html.

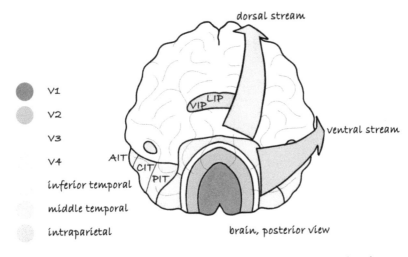

V1

V2

V3

V4

inferior temporal

middle temporal

intraparietal

Extrastriate areas include V2, V3, V4, inferior temporal, middle temporal, and intra-parietal regions. These areas are responsible for various functions involved in further visual processing. Two streams of visual information have been hypothesized: the dorsal and ventral streams, with different objectives in terms of processing.

Q122 **What is V2?**

A122 **V2**, also called visual area 2 or prestriate cortex, is the second major area of the visual cortex, receiving information from V1. V2 is composed of Brodmann areas 18 and 19.

Most V2 neurons have similar properties to neurons in V1. Such properties include: sensitive to orientation, spatial frequency, and color. Functions of V2 include more complex functions such as border ownership, figure-ground segregation, and perception of illusory contours.

Mather, George. "The Visual Cortex." *University of Sussex.* Accessed September 9, 2017. http://www.lifesci.sussex.ac.uk/home/George_Mather/Linked%20Pages/Physiol/Cortex.html.

Q123 **If cytochrome oxidase staining in V1 forms blobs, what does the same stain show in V2?**

A123 Cytochrome oxidase staining of V2 consists of a three-part pattern of thick and thin stripes separated by pale inter-stripes, rather than blobs separated by inter-blobs. Each stripe is thought to have different functions.

Mather, George. "The Visual Cortex." *University of Sussex.* Accessed September 9, 2017. http://www.lifesci.sussex.ac.uk/home/George_Mather/Linked%20Pages/Physiol/ Cortex.html.

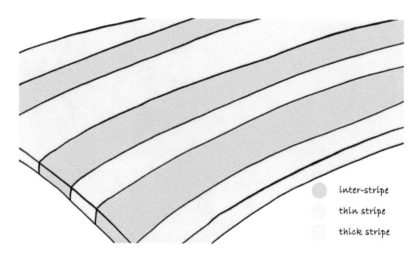

inter-stripe

thin stripe

thick stripe

When V2 is stained with cytochrome oxidase, alternating thin stripes, thick stripes, and inter-stripes appear, each with different functions in visual processing.

Q124 **What are "thin stripes"?**

A124 **Thin stripes** are one region resulting from from cytochrome oxidase staining of the secondary visual cortex. These stripes are thought to function in processing color and are part of the midget/parvocellular pathway. The neurons in the thin stripes receive input from the blobs of V1 and project axons to V4. The color-sensitive cells of the thin stripes have many similarities to the neurons of V1 blobs, with opponent and double-opponent cells. However, unlike blob cells of V1, which generally do not show a preference for orientation, many neurons of V2 are both color- and orientation-selective and are known as color oriented cells. Some subsets of these color oriented cells have new properties that are not seen in V1, such as the color border cell, which responds to a border of a certain orientation between two different colors. Another type of V2 neuron is the color disparity cell, which is selective for color as well as retinal disparity.

Roe, Anna Wang, and Daniel Y. Ts'o. "The Functional Architecture of Area V2 in the Macaque Monkey." In *Extrastriate Cortex in Primates*, 295–333. Cerebral Cortex. Springer, 1997. Accessed September 9, 2017. doi:10.1007/978-1-4757-9625-4_7.

Q125 **What are "thick stripes"?**

A125 **Thick stripes** are another region from cytochrome oxidase staining of the V2 and receive input from the interblobs of V1. Neurons in thick stripes are involved in the perception of depth. Unlike V1, which is organized into ocular dominance columns, where neurons of one column receive monocular input from the same eye, many neurons of V2 receive binocular input. The thick stripes of V2 contain a high proportion of these "depth cells" that are sensitive to retinal disparity. Thick stripes appear to be organized into patches of depth cells of similar properties, such as excitatory or inhibitory as well as the amount of disparity to which the cell is most sensitive.

Roe, Anna Wang, and Daniel Y. Ts'o. "The Functional Architecture of Area V2 in the Macaque Monkey." In *Extrastriate Cortex in Primates*, 295–333. Cerebral Cortex. Springer, 1997. Accessed September 9, 2017. doi:10.1007/978-1-4757-9625-4_7.

Q126 **What is retinal disparity?**

A126 **Retinal disparity** is the way your left and right eye view slightly different images. The two different images "blend" to make a single image. The difference between the two images, known as retinal disparity, is used in depth perception.

An interesting thing to note, is that if an object is closer to you, the retinal disparity is greater, and if the object is farther away the retinal disparity is very little. For example if you put your hand right in front of your eyes, and look at your hand with just your left eye and just your right eye, you see two slightly different images. If you put your hand farther away and try the same thing the two images are less different.

"Retinal Disparity." *Northern Michigan University School of Art & Design*. Accessed September 9, 2017. http://art.nmu.edu/groups/cognates/wiki/1617e/Retinal_Disparity.html.

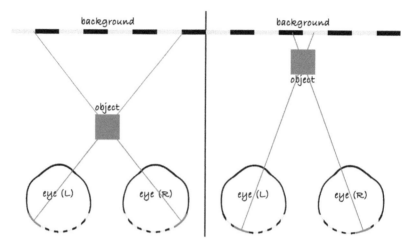

As a result of the slightly different locations of the two eyes, an object is projected to different locations on each retina, a phenomenon known as retinal disparity. The amount of retinal disparity depends on the distance to the object, where farther objects have less disparity.

Q127 **What are "inter-stripes"?**

A127 Pale **inter-stripes** of the V2 are thought to function for the perception of form. V2 inter-stripes receive input from the interblobs of V1. Many neurons of the inter-stripes demonstrate end-inhibition similar to the end-stopped cells of V1. However, unlike the cells of V1, which only respond to real contours, the cells of V2 respond to illusory contours as well.

Roe, Anna Wang, and Daniel Y. Ts'o. "The Functional Architecture of Area V2 in the Macaque Monkey." In *Extrastriate Cortex in Primates*, 295–333. Cerebral Cortex. Springer, 1997. Accessed September 9, 2017. doi:10.1007/978-1-4757-9625-4_7.

Q128 **What are illusory contours?**

A128 **Illusory contours**, also called subjective contours, are a type of visual illusion where contours, or edges, are perceived where there are none, due to the positioning and arrangement of other shapes in an image. A famous example of illusory contours is the **Kanizsa triangle**. In this illusion, there are three Pac-Man shaped figures arranged in a triangle. The viewer perceives a triangle from the edges of the Pac-Man shapes while in actuality there is no triangle. The edges of the created triangle are known as illusory contours.

Simmons, Sheri. "About the Triangle." *The Architectonics of Nature*. Last modified May 25, 1996. Accessed September 9, 2017. https://www.princeton.edu/~freshman/kanizsa.html.

The Kanizsa triangle demonstrates illusory contours, which first to be perceived in V2. In this illusion, a triangle appears where there actually is none. Neurons of V1 with receptive fields covering the edges of the imaginary triangle do not fire; however, the corresponding neurons of V2 do, suggesting that V2 is responsible for the perception of illusory contours like the ones that create the triangle.

Q129 **What is border ownership?**

A129 The brain receives edge information, but in order to identify objects, a contiguous shape must be perceived from the edges. This property of assigning an owner to an edge is known as **border ownership**. Some cells in V2 and other visual areas such as V4 are sensitive not only to the orientation of the edge but also the owner of the border. These cells fire more actively when one side of the edge is an object and the other side is the background, indicating that there is an object on that side of the edge, while the other side is merely background. Assigning owners to edges plays an important role in figure-ground segregation.

Williford, Jonathan R., and Rüdiger von der Heydt. "Border-ownership coding." *Scholarpedia* 8, no. 10 (October 2013). Accessed September 9, 2017. doi:10.4249/scholarpedia.30040.

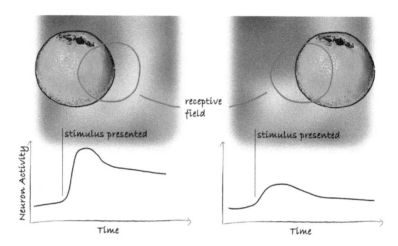

Some neurons of extrastriate areas are sensitive to the "owner" of an edge. If object that contains the edge is on a certain side of the receptive field, the neuron will fire more than it would if the owner is on the other side. This demonstrates that the neuron can distinguish between foreground and background, since the owner of the edge is part of the foreground. This process of assigning owners to borders is the foundation for figure-ground segregation and object identification.

Q130 **What is figure-ground segregation?**

A130 **Figure-ground segregation** is the process of separating the foreground from the background. Distinguishing the figure from the ground is important in the process of object recognition; unclear in border ownership leads to ambiguous figure-ground segregation, which can cause various optical illusions. In the **Rubin vase**, the owner of the edge (shown in red) is ambiguous. This leads to two possible foregrounds: the green area and the blue area. Depending on which is perceived to be the foreground, the viewer can either see a vase or two faces.

Machilsen, Bart. "Figure-ground Organization." *Gestalt ReVision*. Accessed September 9, 2017. http://gestaltrevision.be/en/what-we-do/overview/39-research-areas/mid-level/60-figure-ground-organization.

"Rubin Vase." *New World Encyclopedia*. Accessed September 9, 2017. http://www.newworldencyclopedia.org/entry/Rubin_vase.

The Rubin vase is a visual illusion that works through ambiguous figure-ground segregation. If you perceive the blue to be the foreground and green to be the background, then you see two faces. On the other hand, if you perceive green as the foreground and blue as the background, then you see a vase.

Q131 **What is V3?**

A131 Commonly referred to as the third visual complex, **V3** is a visual area immediately rostral to V2. It has two main regions, the dorsal V3 and the ventral V3. Dorsal V3 has been subdivided into two regions, V3 and V3a. Both receive strong connections from V1 and V2 and project into the posterior parietal areas. They are perceived to be important in processing global motion, or inferential vision. While the V1 focuses on a small visual area, dorsal V3 and V3a respond to extensive fields of vision. The ventral V3 serves a similar purpose, providing an intricate processing of a wide field of vision.

"The Third Visual Complex." Accessed September 9, 2017. https://manumissio. wikispaces.com/%20The%20Third%20Visual%20Complex.

Q132 **What is V4?**

A132 **V4** is a visual area that is comprised of four subdivisions, left and right V4d and left and right V4v (d meaning dorsal and v meaning ventral). V4 receives input from V1 and V2, and has connections with the inferior temporal gyrus, MT, and the dorsal prelunate gyrus. V4 also shows differentiation in firing rates depending on the amount of attention one views an object with, meaning that it responds to attentional modulation. V4 helps process color, contours, and shapes. However, V4 does not recognize faces or geographical areas.

Mather, George. "The Visual Cortex." *University of Sussex*. Accessed September 9, 2017. http://www.lifesci.sussex.ac.uk/home/George_Mather/Linked%20Pages/Physiol/Cortex.html.

Q133   Is V4 the "color area"?

A133   Historically, V4 was known as the "color area", the region in the brain responsible for calculating and processing all color. This is due to claims in the 1970s that all of the neurons in V4 are hue-selective and only fire when a certain **hue** is present in its receptive field. However, these claims have been refuted. Currently, V4 is thought to play a major role in the processing of color; however, it is not solely a color area, as a large portion of its neurons are not hue-selective.

Conway, Bevil R. "Color Vision, Cones, and Color-Coding in the Cortex." *Neuroscientist* 15, no. 3 (2009). doi:10.1177/1073858408331369.

Q134   What are "globs"?

A134   **Globs** are regions in V4 analogous to thin stripes in V2 and blobs in V1. These globs contain neurons that are hue-selective, and respond only to stimulus of a certain color. Differing luminosities of the same color activate these cells equally. Furthermore, glob cells demonstrate lower orientation selectivity than interglob cells. It is possible that globs are arranged in a "**chromotopic map**", where the position of the glob in the cortex corresponds to the hue to which it is most sensitive.

Roe, Anna W., Leonardo Chelazzi, Charles E. Connor, Bevil R. Conway, Ichiro Fujita, Jack L. Gallant, Haidong Lu, and Wim Vanduffel. "Toward a Unified Theory of Visual Area V4." *Neuron* 74, no. 1 (April 12, 2012): 12–29. Accessed September 9, 2017. doi:10.1016/j.neuron.2012.03.011.

Conway, Bevil R., and Doris Y. Tsao. "Color-tuned neurons are spatially clustered according to color preference within alert macaque posterior inferior temporal cortex." *Proc. Natl. Acad. Sci. U.S.A.* 106, no. 42 (October 20, 2009). Accessed September 9, 2017. doi:10.1073/pnas.0810943106.

Q135 **What are "interglobs"?**

A135 **Interglobs** are regions in V4 between globs. Neurons in interglobs do not show hue-selectivity, but do demonstrate higher orientation-selectivity than neurons in globs. These cells are thought to be involved in the processing of shape rather than color.

Roe, Anna W., Leonardo Chelazzi, Charles E. Connor, Bevil R. Conway, Ichiro Fujita, Jack L. Gallant, Haidong Lu, and Wim Vanduffel. "Toward a Unified Theory of Visual Area V4." *Neuron* 74, no. 1 (April 12, 2012): 12–29. Accessed September 9, 2017. doi:10.1016/j.neuron.2012.03.011.

Q136 **What is MT?**

A136 Also called V5 in primates, the medial temporal (**MT**) visual area gets input from V1, V2, and V3, as well as koniocellular parts of the LGN. The MT is important in determining the speed and direction of a moving object, as well as movement of the eye. People with lesions in the MT have been shown to see movement as frames shifting, rather than continuous motion. MT contains neurons that are important in processing the motion of complex objects.

Mather, George. "The Visual Cortex." *University of Sussex*. Accessed September 9, 2017. http://www.lifesci.sussex.ac.uk/home/George_Mather/Linked%20Pages/Physiol/Cortex.html.

Q137 **What is the two-streams hypothesis?**

A137 The **two-stream hypothesis** is a hypothesis describing the processing of visual information in extrastriate areas. According to this model, visual information progresses along two "streams", or pathways: the dorsal and ventral streams.

van Polanen, Vonne, and Marco Davare. "Interactions between dorsal and ventral streams for controlling skilled grasp." *Neuropsychologia* 79 (December 2015): 186–91. Accessed September 9, 2017. doi:10.1016/j.neuropsychologia.2015.07.010.

Q138 **What is the dorsal stream?**

A138 In the two-streams hypothesis, the **dorsal stream**, also known as the parietal stream, "where" stream, or "how" stream, runs from the primary visual cortex in the occipital lobe to various visual regions in the parietal lobe. This pathway is responsible for tracking and guiding movement by creating a detailed visual field, detecting and analyzing movement. The occipital portion is primarily dedicated to pure visual functioning and slowly shifts towards spatial awareness as the pathway approaches the parietal portion, where it is primarily dedicated to spatial awareness. The posterior parietal cortex is integral to coordination of the body and its movements through its two lobules: the lateral intraparietal sulcus (LIP) and the ventral intraparietal sulcus (VIP). The LIP is activated when attention is focused on a stimulus and the VIP integrates visual and somatosensory information.

van Polanen, Vonne, and Marco Davare. "Interactions between dorsal and ventral streams for controlling skilled grasp." *Neuropsychologia* 79 (December 2015): 186–91. Accessed September 9, 2017. doi:10.1016/j.neuropsychologia.2015.07.010.

Q139 **What is the ventral stream?**

A139 In the two-streams hypothesis, the **ventral stream** is the "what" stream of extrastriate visual processing in comparison to the dorsal stream. Starting in V1, the ventral stream travels through V2 and V4, where details of an object are further processed. From V4, the ventral stream goes to the inferior temporal lobe, from the posterior inferotemporal cortex (PIT) to the central inferotemporal cortex (CIT) to the the anterior inferotemporal cortex (AIT), where the ventral stream ends. These are important not just in processing the shape and form of an object, but also the significance of that object. It is recognized that the ventral stream is integral in facial recognition as well as interpreting facial expressions as damage to the ventral stream greatly impairs these systems.

Ungerleider, Leslie G., and James V. Haxby. "'What' and 'where' in the human brain." *Curr. Opin. Neurobiol.* 4, no. 2 (1994): 157–65. Accessed September 9, 2017. doi:10.1016/0959-4388(94)90066-3.

van Polanen, Vonne, and Marco Davare. "Interactions between dorsal and ventral streams for controlling skilled grasp." *Neuropsychologia* 79 (December 2015): 186–91. Accessed September 9, 2017. doi:10.1016/j.neuropsychologia.2015.07.010.

Part II
Development

1 Fetal Development

Q140 **When do the eyes begin development?**

A140 The eyes begin to develop from week 3 to week 8 of pregnancy.

Ort, Victoria, and David Howard. "Development of the Eye." *NYU School of Medicine*. Accessed September 9, 2017. http://education.med.nyu.edu/courses/macrostructure/lectures/lec_images/eye.html.

Q141 **Where are the eyes derived from?**

A141 Human embryos (and those of many animals) divide into three layers: ectoderm, mesoderm, and endoderm. The **ectoderm** is a layer of cells that gives rise to neural and epithelial cells, the **mesoderm** develops into muscles, bones, and blood cells, and the **endoderm** becomes the gastrointestinal and respiratory tracts. In the eyes, the surface ectoderm develops into the non-neuronal parts, for instance the lens, iris, and corneal epithelium, whereas the neuroectoderm differentiates into the neuronal parts denoted as the retina and the optic nerve.

Ort, Victoria, and David Howard. "Development of the Eye." *NYU School of Medicine*. Accessed September 9, 2017. http://education.med.nyu.edu/courses/macrostructure/lectures/lec_images/eye.html.

Q142 **How does the eye develop?**

A142 Eye development begins with the two optic vesicles, which extend from the forebrain, attached by the optic stalks which eventually develop into the optic nerves. The optic vesicles then fold to form the concave structure known as the optic cup, which becomes the retina and inner surface of the eye. The interaction between the optic vesicles and the surface ectoderm cause the latter to divide, forming the lens placode, which eventually develops into the lens.

Ort, Victoria, and David Howard. "Development of the Eye." *NYU School of Medicine.* Accessed September 9, 2017. http://education.med.nyu.edu/courses/macrostructure/lectures/lec_images/eye.html.

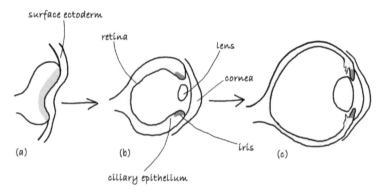

Development of the eye starts from the optic vesicles (a). The optic stalks develop into the optic nerve, while the optic vesicle forms the optic cup containing the retina, ciliary epithelium, and iris (b). The lens and cornea develop from the surface ectoderm. Eventually, a mature eye is formed (c).

Q143 **What chemical signals cause the eye to develop into its mature form?**

A143 A variety of chemicals called transcription factors play a role in the development of the eye. Some examples of these transcription factors are Pax6 and SHH.

Ort, Victoria, and David Howard. "Development of the Eye." *NYU School of Medicine.* Accessed September 9, 2017. http://education.med.nyu.edu/courses/macrostructure/lectures/lec_images/eye.html.

Q144 **What are transcription factors?**

A144 **Transcription factors** are proteins that induce the production of other proteins by causing parts of the genome to become activated.

Cooper, John A. "Transcription factor." *Encyclopædia Britannica.* Accessed September 9, 2017. https://www.britannica.com/science/transcription-factor.

Q145   **What role does Pax6 play in the development of the eye?**

A145   **Pax6** is important for the development of the optic vesicles to form from the neural ectoderm. It can be described as the "master" gene for eye development.

"PAX6 gene." Genetics Home Reference. Accessed September 9, 2017. https://ghr. nlm.nih.gov/gene/PAX6.

Ort, Victoria, and David Howard. "Development of the Eye." *NYU School of Medicine*. Accessed September 9, 2017. http://education.med.nyu.edu/courses/ macrostructure/lectures/lec_images/eye.html.

Q146   **What happens when there is a problem with Pax6?**

A146   Although this is not a clinically significant problem, studies in mice demonstrate that Pax6 mutations cause abnormal development of the eyes. In experiments where the Pax6 gene is completely disabled, known as knockout studies, mice heterozygous for a Pax6 mutation (they have only one working copy of the gene) will have underdeveloped eyes, while mice with homozygous for the Pax6 mutation (they have no working copies of the gene), the eyes will not develop at all.

While knockout studies are artificially induced, natural mutations in the Pax6 gene can also cause eye problems, such as aniridia and Peter's anomaly. In aniridia, the individual lacks one or both irises. In Peter's anomaly, there is leukoma, or clouding of the cornea, as well as abnormalities in the lens, resulting from the cornea, iris, and lens not separating completely during development. Peter's anomaly occurs in about one in 10,000 people.

"Peters anomaly." *Genetics Home Reference*. Accessed September 9, 2017. https://ghr. nlm.nih.gov/condition/peters-anomaly.

"aniridia." *Genetics Home Reference*. Accessed September 9, 2017. https://ghr.nlm. nih.gov/condition/aniridia.

Q147   **What role does SHH play in the development of the eye?**

A147   SHH, which stands for "sonic hedgehog", is a signalling protein involved in the development of many structures in the human body, including the brain and spinal cord. It is required for the splitting of the eyes, forming two eyes rather than a single eye.

Ort, Victoria, and David Howard. "Development of the Eye." *NYU School of Medicine*. Accessed September 9, 2017. http://education.med.nyu.edu/courses/ macrostructure/lectures/lec_images/eye.html.

Q148   What happens when there is a problem with SHH?

A148   In individuals with a mutated SHH gene, only one eye will form in a condition called **cyclopia**.

Ort, Victoria, and David Howard. "Development of the Eye." *NYU School of Medicine*. Accessed September 9, 2017. http://education.med.nyu.edu/courses/ macrostructure/lectures/lec_images/eye.html.

Q149   What is the persistent pupillary membrane?

A149   **Persistent pupillary membrane,** or PPM, is a condition where the fetal membrane persists in tissue spanning across the pupil. This makes the pupil appear segmented, with streaks of the iris running across and dividing it into slices. The strands of fetal membrane could connect to the cornea or lens, but most commonly attaches to the iris. There is not much of an adverse effect nor are there any symptoms if the membrane attaches to the iris, but attachment to the cornea could cause small corneal opacities, or the scarring of the cornea, which could lead to vision impairment and loss. If the membrane attaches to the lens, the patient could experience cataracts. This pupillary membrane originally exists as a source of blood supply to the lens in the fetus, but normally atrophies, or disappears, after a few weeks. If the membrane does not completely atrophy, persistent pupillary membrane occurs.

Vislisel, Jesse. "Persistent pupillary membrane." *EyeRounds*. Accessed September 9, 2017. http://webeye.ophth.uiowa.edu/eyeforum/atlas/pages/persistent-pupillary- membrane.htm.

Gokhale, Varada, and Sumita Agarkar. "Persistent Pupillary Membrane." *N. Engl. J. Med.*, 2017. Accessed September 9, 2017. doi:10.1056/nejmicm1507964.

Q150   What is coloboma?

A150   **Coloboma** is a disease characterized by gaps in certain structures in the eye, such as the iris or the retina. For example, when affecting the iris, the pupil appears keyhole-shaped rather than circular due to missing parts of the iris. Generally, coloboma affecting the iris does not adversely affect vision; however, missing parts of the retina would cause vision loss. Coloboma tends to be associated with other eye disorders, such as myopia, hyperopia, glaucoma, detached retina, and cataracts.

During eye development, a gap known as the **optic fissure**, or the choroid fissure, must close. When the optic fissure does not successfully close, due to environmental or genetic factors, coloboma occurs. Coloboma, and other **congenital** eye disorders, appear to be associated with certain risk factors during pregnancy, such as exposure to alcohol. However, few genes involved in coloboma have been identified.

"Coloboma." *Genetics Home Reference*. Accessed September 9, 2017. https://ghr.nlm. nih.gov/condition/coloboma.

Q151 **How does alcohol during pregnancy affect eye development?**

A151 Drinking alcohol while pregnant can lead to **fetal alcohol syndrome** (FAS), a disease which may lead to mental retardation and learning disability in the newborn. Fetal alcohol syndrome also affects the eyes by reducing visual acuity and causing eye disorders such as farsightedness, nearsightedness, nystagmus, strabismus, and coloboma. Ethanol interferes with glutamate and GABA signaling and causes developing nerve cells in the retina and the visual cortex to undergo apoptosis, a process in which the neuron dies. More than 90% of children affected by fetal alcohol syndrome have some form of eye or visual abnormality.

Tenkova, Tatyana, Chainllie Young, Krikor Dikranian, Joann Labruyere, and John W. Olney. "Ethanol-Induced Apoptosis in the Developing Visual System during Synaptogenesis." *Invest. Ophthalmol. Vis. Sci.* 44 (July 2003). Accessed September 9, 2017. doi:10.1167/iovs.02-0982.

Abdelrahman, Abdelmageed, and Richard Conn. "Eye Abnormalities in Fetal Alcohol Syndrome." *Ulster Medical Society* 78, no. 3 (September 2009). Accessed September 9, 2017. https://www.ncbi.nlm.nih.gov/pmc/articles/PMC2773598/.

2 Newborns and Infants

Q152 **How well can a newborn baby see?**

A152 Historically, it was thought that newborns are almost blind, unable to process any visual information. However, in the mid-20th century, experimentation in infant vision demonstrated that newborns do indeed have to ability to see, albeit not as well as an adult. The visual acuity of a newborn ranges from 20/200 to 20/800. However, as much of the growth and development of the eye takes place in the first year after birth, the vision of the infant reaches adult levels by around six months.

Sokol, Samuel. "Measurement of infant visual acuity from pattern reversal evoked potentials." *Vision Res.* 18, no. 1 (1978). Accessed September 9, 2017. doi:10.1016/0042-6989(78)90074-3.

Q153   **How does being born prematurely affect vision?**

A153   Being born prematurely interrupts the growth of the baby's eyes that occurs during the last 12 weeks of pregnancy. Once born, blood vessels will grow abnormally into the retina. Blood will leak into the eye as a result of these abnormal vessels being quite fragile. Scar tissue that forms after this can pull the retina from the back of the eye, resulting in vision loss. The only way to figure out if your child could have developed a visual defect due to premature birth is an eye exam.

"Facts About Retinopathy of Prematurity (ROP)." *National Eye Institute.* Last modified June 2014. Accessed September 9, 2017. https://nei.nih.gov/health/rop/rop.

Q154   **What is the Fantz preferential looking method?**

A154   The **Fantz method** is a way of performing experiments in infant vision, developed by Robert Fantz in 1961. In this method, an infant is presented with a screen, which is divided into a left section and a right section, each with a different visual stimulus. The side that the infant prefers looking at is recorded.

It has been demonstrated that infants prefer looking at patterned stimuli, such as black and white stripes, rather than a uniform gray region. However, infants only demonstrate this preference if their visual acuity is high enough to distinguish between patterned and uniform regions. If the stripes are sufficiently large, the infant will preferentially look toward the striped side. On the other hand, if the stripes are too fine for the infant to see, both sides will appear uniform gray, and the infant will not demonstrate any preference for either side. By varying the size of the stripes, the visual acuity of the infant can be determined.

Kline, Don. "Preferential Looking." *The Vision and Aging Lab.* Accessed September 9, 2017. http://psych.ucalgary.ca/PACE/VA-Lab/Marcela/Pages/page6.html.

"Visual Perception in Infancy." Accessed September 9, 2017. http://bookbuilder.cast.org/view_print.php?book=28971.

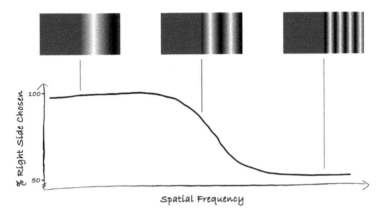

The Fantz preferential method is a method of experimenting with infant vision. The graph shows that the limit of the infant's visual acuity is around the spatial frequency of the middle image, as at higher frequencies (right), the infant cannot distinguish between the textured and uniform sides of the image, resulting in the infant choosing the "correct" side only 50% of the time, due to random chance.

Q155 **What is the habituation method?**

A155 The **habituation method** is another method of experimenting with visual perception in infants. Infants spend more time looking at novel stimuli, or objects and patterns they have not seen before. After spending time around the same stimulus, infants spend progressively less time looking at the stimulus, a process known as habituation. In the habituation method, the infant is presented with the first stimulus for a certain period of time. After the infant becomes use to the stimulus, it is replaced with the second stimulus. If the infant immediately spends more time looking at the second stimulus, then the visual system is capable of distinguishing the two stimuli. If the infant does not spend more time looking at the second stimulus, then he/she likely cannot tell the difference between the two.

Turk-Browne, Nicholas B., Brian J. Scholl, and Marvin M. Chun. "Babies and brains: habituation in infant cognition and functional neuroimaging." *Front. Hum. Neurosci.* 2 (2008). Accessed September 9, 2017. doi:10.3389/neuro.09.016.2008.

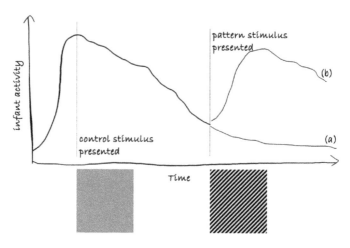

The habituation is another method of experimenting with infant vision. First, a uniform image is presented. At the sight of this new stimulus, infant activity increases, but eventually drops back down as the infant becomes habituated to the stimulus. When the uniform control stimulus is switched with a patterned stimulus, there are two possibilities. If the pattern is beyond the visual acuity of the infant, then the infant cannot tell the difference and does not realize there is a new stimulus, leading to the same low level of activity (a). If the pattern within the visual abilities of the infant, then he/she increases in activity at the sight of a new stimulus (b). The experiment can be repeated with various patterned stimuli to determine the limit of infant visual acuity.

Q156 **How does color vision develop in infants?**

A156 Visual experience with different colors in infants appears to be essential to proper visual development. In one experiment, infant monkeys were raised for about a year illuminated only by monochromatic light, suppressing the range of colors visible to them. They were then moved to a room with normal lighting. The monkeys were presented first with a task testing only brightness, which they passed. Next, they were presented with another task that tested the ability to discriminate between colors. Monkeys raised in a monochromatic environment demonstrated poor color discrimination, grouping colors such as red and blue together as "similar", while humans and monkeys raised in normal environments did not. Clearly, visual experience with a wide range of colors is necessary for proper visual development in infants during this critical period.

Sugita, Yoichi. "Experience in Early Infancy Is Indispensable for Color Perception." *Curr. Biol.* 14, no. 14 (July 2004): 1267–71. Accessed September 9, 2017. doi:10.1016/j.cub.2004.07.020.

Q157 **What is the effect of visual deprivation in infants?**

A157 Visual deprivation in infants would cause the loss of function in the eye(s) deprived of light, which can range from lowered visual acuity to complete loss of vision in the eye. Visual deprivation is especially detrimental during infancy because the visual system critical period for humans occurs from birth to 6 months of age, during which visual stimulus is essential to the proper development of the visual system.

Hubel and Wiesel performed several experiments to determine the exact nature of the effect of visual deprivation on the developing brain. They sewed the eyelids of one eye of kittens shut so that it would receive no light. Afterward, they discovered that while neurons in the lateral geniculate nucleus developed mostly normally, many cells in the visual cortex did not respond to light normally. They then conducted another experiment, where they covered one eye of kittens with a translucent cover so that it did not completely block out light, but prevented the kitten from seeing any shapes or form. Again, the visual systems did not develop correctly. Thus, visual experience with form perception during infancy is critical to proper visual development.

This is applicable to humans, as some children are born with cataracts, a disease caused by the clouding of the lens. These cataracts may impair vision during the critical period of visual development, thus causing permanent visual impairment, and should be treated quickly.

Cooper, Jeffrey. "What is Strabismus?" *Optometrists Network*. Accessed September 9, 2017. http://www.strabismus.org/vision_therapy_for_strabismus.html.

Domenici, L., N. Berardi, G. Carmignoto, G. Vantini, and L. Maffei. "Nerve growth factor prevents the amblyopic effects of monocular deprivation." *Proc. Natl. Acad. Sci. U.S.A.* 88, no. 19 (1991). Accessed September 9, 2017. doi:10.1073/pnas.88.19.8811.

Hubel, D. H., and T. N. Wiesel. "Effects of monocular deprivation in kittens." *Naunyn Schmiedebergs Arch Exp Pathol Pharmakol.*, August 19, 1964. Accessed

September 9, 2017. http://hubel.med.harvard.edu/papers/HubelWiesel1964Naunyn SchmiedebergsArchExpPatholPharmakol.pdf.

Q158 **What is depth perception?**

A158 **Depth perception** is the ability to use binocular vision, also known as stereopsis, along with other visual cues, to determine the distance to an object.

Wilkinson, C. P. "Treatment for Loss of Depth Perception." *American Academy of Ophthalmology*. Last modified September 28, 2015. Accessed September 9, 2017. https://www.aao.org/eye-health/ask-ophthalmologist-q/is-there-treatment-loss-of-depth-perception.

Q159 **What is stereopsis?**

A159 **Stereopsis**, or binocular vision, is where visual information from both eyes is used to determine the distances to objects. Despite the fact that distances can be approximately judged using a single eye, binocular vision is essential to accurately gauging distances, especially at close range. To prove this point, try threading a needle with only one eye.

However, the accuracy of stereoscopic vision for depth perception decreases as objects get further, as the amount of separation between the two eyes becomes increasingly negligible compared to the actual distance to the object.

Shugarman, Richard G. "What Is Stereopsis?" *American Academy of Ophthalmology*. Last modified March 9, 2013. Accessed September 9, 2017. https://www.aao.org/eye-health/ask-ophthalmologist-q/stereopsis.

Q160 **What are some visual cues besides stereopsis that are used in depth perception?**

A160 Other visual cues that are used to determine the distance to an object include size and motion parallax. Closer objects appear to be larger than farther objects of the same size. Similarly, in **motion parallax**, closer objects move faster across the visual field than farther objects travelling at the same velocity. Both of these are used in depth perception.

Krantz, John H. "Motion Parallax." *PsychScholar*. Accessed September 9, 2017. http://psych.hanover.edu/Krantz/MotionParallax/MotionParallax.html.

Q161 **How do 3D glasses work?**

A161 Most 3D glasses in the movie theatres use polarization to give a sense of depth to it viewers. To understand how these 3D glasses work, we must learn some basic optics first.

Polarization refers to how a light wave is oriented. Normal light consists of light oriented in all directions. A 3D movie is filmed with two cameras, separated by approximately the same distance as between the two eyes. When presented, two images from the cameras that are superimposed, each image projected with light only at a certain orientation. The 3D glasses of the viewer has two lenses, one for each eye. The left lens only allows polarized light of the same orientation as the first image to enter the eye, while the right lens only allows polarized light of the same orientation as the second image to enter the eye. Thus, in each eye, a different image is received from the same screen. With the different images, retinal disparity allows depth to be perceived, hence perceiving a 3D image.

Kevin, Rio. "How It Works: The Evolution of 3D Glasses and 3D Technology." *Journal of Young Investigators*. April 2007. Accessed September 07, 2017. http://www.jyi.org/issue/how-it-works-the-evolution-of-3d-glasses-and-3d-technology/.

Q162 **When does a newborn acquire depth perception?**

A162 It takes about five months of living and seeing for a newborn to learn to judge distances.

"Infant Vision: Birth to 24 Months of Age." *American Optometric Association.* Accessed September 07, 2017. https://www.aoa.org/patients-and-public/good-vision-throughout-life/childrens-vision/infant-vision-birth-to-24-months-of-age.

Q163 **How does a newborn acquire depth perception?**

A163 A large part of gaining depth perception is in learning from experience. An example is when infants learn size constancy, where objects appear smaller when far away, but remain the same size in reality. Furthermore, as newborns age, they gain more experience in using both eyes and combining the image from each eye into a single three-dimensional image. At the same time, the brain also develops pathways to analyze and combine the images formed by the eyes to determine depth and distance.

It is interesting to note that the development of depth perception occurs around the same time as the infant begins to crawl. Depth perception is necessary for infants to move about safely, as the infant would need to be able to judge the distance to a potentially dangerous location, like a cliff. In 1960, Eleanor J. Gibson and R.D. Walk created an experiment in which infants were placed on a "**visual cliff**" apparatus to study the development of sensory processing in infants.

Rodkey, Elissa N. "The woman behind the visual cliff." *Monitor On Psychology* 42, no. 7 (2011): 30. http://www.apa.org/monitor/2011/07-08/gibson.aspx.

3 Children

Q164 **What is a conjunctivitis, or pink eye?**

A164 **Conjunctivitis**, or pink eye, is a condition commonly found in newborns and children, as well as adults, caused by certain pathogens, allergies, foreign objects, or chemicals in the eye. Common symptoms of the eye include redness, swelling, burning, tearing, crust-formation, and light-sensitivity.

Viral conjunctivitis produces watery discharge in either one or both of the eyes. On the other hand, bacterial conjunctivitis, which is more common in children, produces a discharge of a yellow-green color. Both types are usually associated with respiratory infections such as sore throats.

Allergic conjunctivitis is caused by allergens, mainly pollen, and usually affects both eyes. Allergens cause antibodies to be released, leading to an inflammatory reaction in the eyes. This type of pink eye is usually associated with other allergic symptoms such as itching, tearing, and watery nasal discharge.

Conjunctivitis can also be caused by irritation from chemicals or a foreign object, leading to watery eyes and mucous discharge. In this case, the foreign object or chemical may need to be removed in order for the eye to clean up.

"Pink eye—Know when to treat versus wait it out." *Mayo Clinic*. June 20, 2017. Accessed September 10, 2017. http://www.mayoclinic.org/diseases-conditions/pink-eye/basics/causes/con-20022732.

"Pinkeye (Conjunctivitis) Slideshow: Causes, Symptoms, & Treatments" *WebMD*. Accessed September 10, 2017. http://www.webmd.com/eye-health/ss/slideshow-pinkeye.

Q165 **What is refractive power?**

A165 **Refractive power**, also known as optical power, quantifies the curvature of the cornea in diopters. This measures the degree to which any optical system converges or diverges light. Optical power and focal length are correlated inversely, meaning that the greater the focal length is, the lower the optical power, and vice versa. Specifically, refractive power and focal length are related through the following relationship:

$$D = \frac{1}{f}$$

where D is the refractive power in diopters and f is the focal length in meters.

Refractive power is important in vision it is a measure of how much light bends. Both the cornea and lens refract light, and imperfections in the refractive powers of the two structures leads to blurry vision in several different types of eye diseases.

"Shape, Curvature, and Power." *American Academy of Ophthalmology*. Accessed September 10, 2017. https://www.aao.org/bcscsnippetdetail.aspx?id=99d3bbfa-cd4e-4615-9bd9-8119c0d7bc7a.

Q166 **What is myopia?**

A166 **Myopia**, also known as nearsightedness, is a common condition of the eye that affects over a quarter of the population, and up to 90% in some Asian countries. Myopia is characterized by difficulty viewing distant objects clearly, but normal vision while viewing close objects. Fortunately, this condition is easily manageable by using glasses, contact lenses, or laser eye surgery.

Myopia is caused by one of two things. Either the eyeball is too long, or the the cornea is curved more than usual. Both conditions create a refractive error, making the light entering the eye bend incorrectly, focusing the light in front of the fovea instead of perfectly on the fovea. Some common symptoms of patients with myopia include blurry vision when viewing far objects, squinting, fatigue, and eye strains that may cause headaches.

Myopia is influenced by both genetic and environmental factors. Patients commonly get this disease as children and it may worsen as they age. Some environmental factors include education levels, and time spent outdoors. Myopia is usually more common in college students compared to less educated populations. Furthermore, spending more time outside appears to reduce the risk of developing myopia. To diagnose myopia, standard eye exams are given by eye doctors.

Juvenile (or school) myopia, early onset myopia occurring in ages from 9–11 years old, accounts for 60% of all cases. Genetic and environmental factors contribute to the development of juvenile myopia.

Grayson, C. E. "Myopia: Learn the Definition, Causes and Control." *MedicineNet.* October 2004. Accessed September 10, 2017. http://www.medicinenet.com/myopia/article.htm.

Myrowitz, Elliott H. "Juvenile myopia progression, risk factors and interventions." *Saudi J. Ophthalmol.* 26, no. 3 (2012): 293–97. doi:10.1016/j.sjopt.2011.03.002.

Goldschmidt, E., and N. Jacobsen. "Genetic and environmental effects on myopia development and progression." *Eye.* 28, no. 2 (2013): 126–33. doi:10.1038/eye.2013.254.

Q167 **What type of lens is used to treat myopia?**

A167 Myopia, or nearsightedness, is usually treated with divergent lenses. Furthermore, some studies have suggested that multifocal contact lenses can slow the progression of myopia and children. Studies have supported that conventional single vision lenses were less effective compared to bifocal and multifocal lenses. Other methods of slowing myopia progression that have been investigated include the use of cycloplegic drugs, such as atropine. However, the effects of atropine on myopia progression are limited and possibly only beneficial in the short-term.

Heiting, Gary. "Nearly Half of the Global Population May Be Nearsighted by 2050." *All About Vision*. Accessed September 10, 2017. http://www.allaboutvision.com/parents/myopia.htm.

"Nearsightedness and its Correction." *The Physics Classroom*. Accessed September 10, 2017. http://www.physicsclassroom.com/class/refrn/Lesson-6/Nearsightedness-and-its-Correction.

Q168 **What is hyperopia?**

A168 Commonly known as farsightedness, **hyperopia** is a condition similar to myopia, except the light is focused behind the retina instead of in front of it. This results in blurry images for near objects while maintaining normal acuity with far objects. In more extreme cases, farsightedness can lead to the lack of an ability to see both far and near objects. This occurs most commonly in children, but affects adults as well. Common symptoms of hyperopia include eyestrain and frequent squinting. A common cause of hyperopia is an error in corneal shape causing the light to focus behind the retina rather than on the fovea. Other causes include weak ciliary muscles, resulting in the lack of lens contractility, diabetes, and blood vessel problems in the retina. The most basic treatment of farsightedness is wearing **contact lenses** or glasses. To eliminate the issue, however, there are procedures that remove part of a corneal surface and also laser eye surgery could solve the specific issue that is causing farsightedness.

"Facts About Hyperopia." *National Eye Institute*. July 01, 2016. Accessed September 10, 2017. https://nei.nih.gov/health/errors/hyperopia.

Q169 **What type of lens is used to treat hyperopia?**

A169 There are three types of lenses that can help correct hyperopia: single, astigmatic, and multifocal lens. Single lens assist in vision for all distances and astigmatic lens correct both hyperopia and astigmatism. In contrary to single lens, **multifocal lens** uses different prescriptions on one lens, allowing different lens to be used for certain distances. Another type of lens to correct hyperopia are progressive lens. These lens are essentially bifocals lens with a more smooth transition between the different prescription of the same lens. This eliminates visible dividing lines.

Bailey, Gretchyn. "Hyperopia (Farsightedness)." *All About Vision.* Accessed September 21, 2017. http://www.allaboutvision.com/conditions/hyperopia.htm.

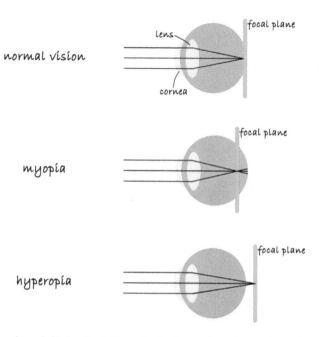

In a normal eye, light is refracted to a focal point on the retina so that a clear image is produced. In myopia, the focal plane is in front of the retina, leading to increasingly blurry images as the object being viewed gets farther. In hyperopia, the focal plane is behind the retina, leading to an increasingly blurry image as the object being viewed gets closer.

Q170 **What is astigmatism?**

A170 **Astigmatism** is a condition where the cornea has an abnormal curvature, causing problems with focusing light on the retina. This is very common in growing children; one study found that 28.4% of all children ages 5 to 17 in America have astigmatism. Furthermore, one third of all people with glasses have astigmatism in at least one eye. If astigmatism is left untreated, it can lead to permanent visual disabilities. Unlike in myopia and hyperopia, in astigmatism, the light is not refracted the same amount in different directions. This is usually caused by the cornea not being equally curved in all directions (imagine the surface of an American football as opposed to the surface of a basketball). In most cases astigmatism is present since birth. However, it can also be a result of eye injury, through trauma or eye surgery. Drinking alcohol during pregnancy can also increase risk of the baby developing astigmatism. Astigmatism can be corrected with glasses which counterbalance the corneal deformity, or with corrective corneal surgery like LASIK or PRK.

"Glasses for Children." *American Association for Pediatric Ophthalmology and Strabismus*. Accessed September 10, 2017. https://www.aapos.org/terms/conditions/54.

"What Percentage of the Population Wears Glasses?" *Glasses Crafter*. Accessed September 10, 2017. http://glassescrafter.com/information/percentage-population-wears-glasses.html.

In astigmatism, a directionally blurred image is produced instead of a uniformly blurred image.

Q171 **What are common instruments used in an eye exam?**

A171 While visiting your eye doctor, they may use a variety of tools to examine your eyes:

When checking for myopia or hyperopia, a **phoropter** is used, which has multiple lenses through which an eye chart is viewed. The lens that provides the clearest image is used to create a prescription for correcting refractive error. An **autorefractor** is a machine that does the same thing as the phoropter but is completely automated and requires no patient feedback to determine refractive error.

A **slit lamp** is a microscope used to examine your eyes. This instrument uses a high-intensity light source that shines a thin sheet of light into the eye. This instrument can examine both the anterior and posterior segments of the eye.

Additionally, the **ophthalmoscope** is another instrument imperative for the diagnosis of eye health. There are many types of ophthalmoscopes, including direct, indirect, monocular, and binocular variants. Direct ophthalmoscopy is commonly used in regular checkups, while indirect ophthalmoscopy has a higher field of view and can detect more subtle lesions.

A visual field machine is used to diagnose any deficiencies in the retina or visual pathway from the eye to the brain. It helps diagnose vision field loss by instructing the patient press a clicker when they see a certain light pattern across a screen. This helps define the patient's field of vision.

To measure the patient's intraocular pressure, which is the pressure within the eye, a **tonometer** is used. The preferred gold standard of diagnosing intraocular pressure is the Goldmann tonometer.

Finally, the **keratometer** and **corneal topographer** measure the curve of the cornea, providing important information for degree of refractive error and diagnosing keratoconus, a disease affecting the cornea. A corneal topographer offers higher resolution of the cornea than the keratometer.

"Slit-lamp exam." *MedlinePlus*. Accessed September 10, 2017. https://medlineplus.gov/ency/article/003880.htm.

Bickford, L. "Instruments Commonly Used For Examination of the Eye." *Ophthalmic Examination Instruments*. Accessed September 10, 2017. http://see.eyecarecontacts.com/instruments.html.

Q172 **What are common tests used in eye exams?**

A172 There are a plethora of eye exams that ophthalmologists use to diagnose eye conditions of the eye ranging from simple to complex tests. Some common exams include the following:

Visual acuity tests examine the sharpness of the patient's vision usually through Snellen charts. The patient is instructed to read an eye chart with letters of different sizes. These charts help indicate the effect of distance on the patient's vision.

A Snellen chart is commonly used to determine visual acuity.

Color blindness tests help diagnose any color deficiencies present in the eye. These tests usually consist of many colorful dots that arranged to usually make a number. If the patient has trouble figuring out the number, the patient will most likely have a form of color blindness.

The **cover test** is another common test which helps diagnose strabismus (crossed eye) or amblyopia (lazy eye). During this test, the doctor will cover one of the patient's eye and instruct the patient to use that eye to focus on a certain nearby object. Then the doctor will cover the next eye, and instruct the patient to focus on the same object. If either one or both of the eyes cannot focus on the object, the doctor may diagnose an eye condition such as amblyopia or strabismus.

Ocular motility testing is a simple test to determine if the patient can move their eyes normally. The test is performed either when a doctor moves an object to see if the eye can follow it or when the doctor holds up to objects and ask the patient to constantly shift focus from one object to another.

Stereopsis testing is used to test the patient's depth perception. This test is usually performed by the doctor instructing the patient to look at a booklet while wearing 3D glasses. The patient must determine which of the four circles presented in the book seem closer to them. If the patient is able to distinguish which circle is "closer" to them, then they have normal depth perception.

A **retinoscopy** is another eye test used to determine which prescription of eyeglasses would be optimal for the patient. This test is performed in a dim room while the patient focuses on a large object, such as the large E on an eye chart. The doctor shines a light on the patient's eye and uses a machine to constantly switch the lens in front of the eye. The lens that is optimal for the patient's vision, helps determine the prescription needed for the patient. The doctor is able to tell if the lens is the best choice based on how the eye reflects off the patient's eye.

Heiting, Gary, and Jennifer Palombi. "What To Expect During A Comprehensive Eye Exam." *All About Vision*. Accessed September 10, 2017. http://www.allaboutvision.com/eye-exam/expect.htm.

Q173 **What is 20/20 vision?**

A173 Normal vision is often referred to as 20/20 vision. When doctors check your vision, they usually use a **Snellen chart**, containing letters with multiple rows decreasing in size. The patient stands twenty feet away from the chart, and if the patient is able to correctly read the letters from a certain row he/she is thought to have normal vision. The 20/20 vision means that when the patient is 20 feet away, he/she can see details that people with normal vision can see twenty feet away. If you have 20/30 vision, that means that you see details comfortably twenty feet away, whereas people with normal vision can see details of the same thing thirty feet away. In rare cases, a patient may have

20/10 vision. This means that he/she can see details twenty feet away when it would take most people 10 feet to see the same details. This vision is above average and you can see more clearly than others. If someone has 20/200 vision, then he/she needs 20 feet to see something that people with regular vision can see from 200 feet away. Note that 20/20 vision is not in any way "perfect vision".

"What is a: Snellen Chart." *EyeGlass Guide*. Accessed September 10, 2017. http://www.eyeglassguide.com/my-visit/vision-testing/snellen-chart.aspx.

Q174 **What is an eyeglass prescription and how is it notated?**

A174 Eyeglass prescriptions are instructions from eye doctors that provides the values of each parameter to correct appropriate lenses for individual patients, and it is required in the U.S. for doctors to give a copy of their eye prescriptions if the patients requires corrective lenses. There are two types of prescriptions: single vision (SV) and multifocal (MF).

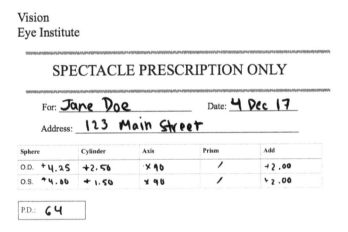

Vision
Eye Institute

SPECTACLE PRESCRIPTION ONLY

For: Jane Doe   Date: 4 Dec 17
Address: 123 Main Street

	Sphere	Cylinder	Axis	Prism	Add
O.D.	+4.25	+2.50	×90	/	+2.00
O.S.	+4.00	+1.50	×90	/	+2.00

P.D.: 64

Eyeglass prescriptions are used to specify the parameters necessary for glasses to correct refractive error.

Multifocal lenses contain multiple lens- each lens are used for different distances. For example, there is usually one near (located on the bottom), one distance (on top), and one intermediate lens (in between). On the other hand, single vision lenses have one lens with a constant power throughout the lens.

In these prescriptions there is information for both eyes. Oculus Dexter (OD) indicates the right eye, and Oculus Sinister (OS) indicates the left eye. SPH, or sphere, is the amount of refractive power (measured in diopters) required to correct the patient's vision, such as for myopia and hyperopia. The reason why SPH treats myopia and hyperopia is because in these eye conditions, there is a blur in all directions, and spherical lenses magnify in all directions, as opposed to cylindrical lenses. The + and—sign indicate whether a plus or a minus lens would be used. **CYL**, or cylinder, is the refractive power needed to cure astigmatism (condition where the eye that creates blurry vision). In astigmatism, the cornea has a higher refractive power in one direction (e.g. horizontally) as opposed to another direction (e.g. vertically). To correct this, lenses with cylindrical correction have different refractive powers in different directions. A positive CYL number indicates farsighted astigmatism, while a negative CYL number indicates nearsighted astigmatism. Axis (X) is a measurement of orientation of the cylindrical correction, to correct for astigmatism with axes that are not perfectly vertical and horizontal. It can range from 1–180°. An empty CYL section indicates no astigmatism. The Add section indicates the refractive power needed to correct for presbyopia at the bottom of the lens. If this part is empty, the patient does not need multifocal lenses.

Heiting, Gary. "How To Read Your Eyeglass Prescription." *All About Vision.* Accessed September 10, 2017. http://www.allaboutvision.com/eyeglasses/eyeglass-prescription.htm.

Heiting, Gary. "Bifocals & Trifocals Solutions." *All About Vision.* Accessed September 10, 2017. http://www.allaboutvision.com/lenses/multifocal.htm.

Q175 **What are LASIK and PRK, and what is the difference between the two?**

A175 LASIK is an abbreviation for **laser-assisted in situ keratomileusis,** which is the most common type of laser eye surgery, and is used to correct myopia, hyperopia, and astigmatism. Similarly, **photorefractive keratectomy** (PRK) is another method of laser eye surgery that was commonly used before LASIK was invented to correct visual problems. In both methods, a laser is used to reshape the cornea so that it focuses light more accurately onto the retina. Both techniques have the same overall effect, but achieve it in different ways.

In LASIK, first the surgeon will make an incision in the cornea, creating a flap of tissue. This tissue is then lifted, allowing the laser to carefully reshape the cornea, hence repairing any imperfections. The flap is then reattached and heals within a few days.

PRK, on the other hand, does not create a flap of corneal tissue. Rather in PRK the whole corneal epithelium is simply removed, which allows the laser to correct the deeper layers of the cornea. A few days are then required for the epithelium to regenerate and for vision to stabilize. Despite the dominance of LASIK, PRK can be a better choice for certain people who have thin corneas or chronically dry eyes.

One reason LASIK is commonly preferred is due to the fact that after surgery it is faster and less painful for the cornea to heal. Although both LASIK and PRK patients are prescribed eye drops after the surgery to alleviate pain, PRK requires days for the epithelium to regrow, compared to the few hours of recovery time required in LASIK. Both techniques are quite safe with high success rates, and the decision between on which option is best is decided by an eye specialist or surgeon, and usually comes down to the parameters of the eye.

"Medical Definition of LASIK." *MedicineNet*. Accessed September 10, 2017. http://www.medicinenet.com/script/main/art.asp?articlekey=7849.

"PRK vs. LASIK. Why is PRK right for you." *LASIK.com*. March 1, 2014. Accessed September 10, 2017. https://www.lasik.com/articles/lasik-prk-difference/.

Q176 **What are night contacts and how do they work?**

A176 Night contacts are a type of eye contacts that some people wear overnight. These contacts are also known as orthokeratology (or ortho-k) and are prescribed to patients who have nearsightedness or blurred vision. They are gas permeable contacts that are to be worn overnight to reshape the cornea. The cornea is essential for focusing light onto the retina through refraction. The contacts reshape the cornea to properly focus light. After wearing them overnight, the following day, patients are able to see clearly. Patients should wear these contacts every night for optimal results.

The efficacy of these night contacts over extended periods of time is still uncertain. Long-term studies on ortho-K suggest that night contacts are not a permanent solution for myopia, and there is a higher risk of contact-related side effects from prolonged use. On the other hand, other studies support the safety and efficacy of orthokeratology..

Campbell, E. J. "Orthokeratology: An Update." *Optom. Vis. Perform.* 1, no. 1 (2013): 11–18. doi:10.1111/j.1444-0938.2006.00044.x.

Liu, Yue M., and Peiying Xie. "The Safety of Orthokeratology—A Systematic Review." *Eye Contact Lens.* 42, no. 1 (2016): 35–42. Accessed September 10, 2017. doi: 10.1097/ICL.0000000000000219.

Heiting, Gary. "Orthokeratology: Reshaping Your Eyes With Contact Lenses." *All About Vision.* September 2016. Accessed September 10, 2017. http://www.allabout vision.com/contacts/orthok.htm.

Q177 **What are childhood cataracts?**

A177 **Cataracts** are commonly found in older people, however some children and newborns have cataracts. 1 in 250 children will develop cataracts either before birth or during childhood. Cataracts are an eye condition where the lens of the eye becomes cloudy, causing blurred vision. Cataracts can be in either in a single eye or both eyes. Some children are born with cataracts, which are known as congenital cataracts. Often, cataracts in children are caused by genetics or simply by eye trauma.

To treat cataracts, many hospitals recommend surgery, as cataracts may gradually develop blindness or amblyopia (lazy eye). Surgery is usually painless and provides very little discomfort to children. Once the cataract is removed through surgery it is imperative that focusing power must be restored. To restore the focusing power of the lenses, either contact lenses, glasses, or intraocular lenses are recommended. Intraocular lenses are artificial lenses implanted to replace the removed lens. Additionally, unilateral cataracts (cataracts in only one eye) may cause amblyopia. To treat amblyopia using an eye patch on the dominant eye to stimulate vision in the lazy eye is recommended. After treatment, an "after cataract" may form. After cataracts are cataracts that form after surgery for the original one. After cataracts usually would require laser surgery to correct the lens.

The prognosis of childhood cataracts vary. If the cataract was acquired then the prognosis is better than congenital cataracts, as in acquired cataracts visual development in the brain already occurred.

"Cataracts in Children Information & More." *Cleveland Clinic.* March 22, 2015. Accessed September 10, 2017. https://my.clevelandclinic.org/health/articles/pediatric-cataracts.

VanderVeen, D. K. "Cataracts in Children." *Boston Children's Hospital.* Accessed September 10, 2017. http://www.childrenshospital.org/conditions-and-treatments/conditions/cataracts.

Q178 What is amblyopia?

A178 Amblyopia, also known as lazy eyes, is an eye condition where there is a loss of vision in one or both eyes which is a result of abnormal visual development. Around 1 in 40 children have this eye condition.

There are several possible causes for amblyopia. The most common cause is strabismus, or misaligned eyes. In this case, because the eyes are not pointed toward the same direction, the brain chooses the eye with the more dominant image, which causes the optic nerve of that eye to become more developed. The eye not being used then becomes less developed. Another cause of amblyopia is if the two eyes have significantly different prescriptions, which is known as **anisometropia**, or if one eye has a cataract and the other does not. As a result, the brain is unable to combine the two images, and the clearer one is used while the blurrier one is ignored. This can lead to less visual development in the consistently ignored eye. Furthermore, due to the usage of only one eye, amblyopia causes a decrease in depth perception.

It is possible to treat some forms of amblyopia. Treatment usually consists of wearing an eye patch a couple hours a day over the dominant eye, which would help train the lazy eye. This is known as occlusion therapy. Surgery on the eye muscles may be necessary in severe cases. Treating amblyopia is effective, but a gradual process that will take months to work on. The prognosis is best in patients diagnosed early. Thus, it is important to have eye exams, especially in young children.

Krucik, G. "What Causes Lazy Eye?." *Healthline.* Accessed September 10, 2017. http://www.healthline.com/symptom/lazy-eye.

"Lazy eye (amblyopia)." *NHS Choices.* June 16, 2016. Accessed September 10, 2017. http://www.nhs.uk/conditions/Lazy-eye/Pages/Introduction.aspx.

"Amblyopia." *The Free Dictionary.* Accessed September 10, 2017. http://medical-dictionary.thefreedictionary.com/amblyopia.

Q179   What causes strabismus?

A179   Strabismus, or crossed eyes, is often caused by poor neuromuscular control of the eyes by weak or palsied **cranial nerves** responsible for eye movement. The nerves controlling eye movement are cranial nerves III (oculomotor), IV (trochlear), and VI (abducens).

Strabismus can be caused by a multitude of problems, whether it is the muscles, nerves, or part of the brain that controls eye movement. Risk factors include a family history of strabismus, refractive error (such as hyperopia), or other medical symptoms including Down syndrome and cerebral palsy.

Strabismus may be treated by eyeglasses or contacts. Special lenses called prism lenses may also be used. These lens alter the light entering, requiring the eye to do less work. Another option for treating strabismus is through vision therapy. Vision therapy consists of a structured program of visual activities that improve eye movement, improving the connection between the eye and the brain. In some cases, eye surgery may be necessary. This surgery balances the muscle of the eye by elongating or shortening it.

"Strabismus (Crossed Eyes)." *American Optometric Association.* Accessed September 10, 2017. https://www.aoa.org/patients-and-public/eye-and-vision-problems/glossary-of-eye-and-vision-conditions/strabismus?sso=y.

Part III
Aging

Q180 How does aging affect vision in general?

A180 Overall, as humans grow older is natural for our vision to decrease in performance. Presbyopia and cataracts are common eye conditions for the elderly. Both conditions are easily curable through reading glasses and cataract surgery. However, some age-related eye diseases are more severe, such as glaucoma, macular degeneration, and diabetic retinopathy.

Garrity, J., W. MacMillan, and B. MacMillan. "Effects of Aging on the Eyes—Eye Disorders." Merck Manuals Consumer Version. Accessed September 10, 2017. http://www.merckmanuals.com/home/eye-disorders/biology-of-the-eyes/effects-of-aging-on-the-eyes.

Q181 How does the performance of the visual system deteriorate with age?

A181 One of the methods to quantify quality of the visual system is through a **contrast sensitivity function** (CSF). A CSF is a graph showing the threshold contrast level as a function of the spatial frequency of the image. Spatial frequency reflects visual acuity, where low spatial frequency is low in detail and high spatial frequency is extremely detailed and fine. At any given spatial frequency, there is a threshold contrast level of the image where at higher contrast it is visible to the viewer and at lower contrast it is invisible.

Studies conducted by measuring the contrast sensitivity functions of older adults have suggested that the majority of the decrease in sensitivity, at least under brightly lit (photopic) conditions, is a result of optical issues, like refractive errors, presbyopia, and clouding of the lens. Other recent studies, however, indicate that neuronal damages, like a decrease in ganglion cell density, also contributes significantly to age-related vision deterioration.

Owsley, C. "Aging and Vision." *Vision Res.* 51, no. 13 (2011): 1610–622. doi:10.1016/j.visres.2010.10.020.

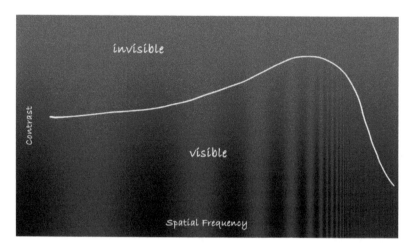

The contrast sensitivity function (white) shows the limit of visual acuity. Above the line, the contrast is not enough for the detail to be visible. Below the line, the contrast is enough for the pattern to be seen. The amount of contrast for a pattern to be visible depends on the spatial frequency, or amount of detail, of the pattern, which varies on the horizontal axis in the graph.

Q182 **What are the effects of aging on the muscles of the eye?**

A182 There are a few notable changes to the eyelid muscles of an individual as a result of aging. The muscles used to close one's eyes become weaker. This may cause a condition known as **ectropion**, where the lower eyelid **tends** to turn outward, exposing the inner eyelid. Another possibility as a result of the looseness of the muscles is a condition known as **entropion**, where the lower eyelid turns inward, causing the eyelid to rub against the eyeball.

Another group of muscles in the eye that weaken with age are the iris muscles that control the size of the pupil. This results in the pupil responding more slowly to different light conditions. This decrease in speed of dark adaptation may result in elders to be dazzled when entering a bright environment right after a dark one or vice versa, for their pupil is slow to adjust.

Garrity, J., W. MacMillan, and B. MacMillan. "Effects of Aging on the Eyes—Eye Disorders." *Merck Manuals Consumer Version.* Accessed September 10, 2017. http://www.merckmanuals.com/home/eye-disorders/biology-of-the-eyes/effects-of-aging-on-the-eyes.

Q183 **How does aging affect the sclera?**

A183 Aging can cause dehydration in the sclera. This dehydration results in calcium deposition, making the sclera more rigid, and may cause scleral calcification. Furthermore, aging may cause yellowing in the sclera, changing the external appearance of the eye, and translucent spots may begin to form throughout the sclera.

Mccormack, John. "The aging individual: physical and psychological perspectives (2nd edition)." *Australas. J. Ageing.* 24, no. 4 (2005): 218–19. Accessed September 10, 2017. doi:10.1111/j.1741-6612.2005.00127.x.

Grytz, R., M. A. Fazio, M. J. Girard, V. Libertiaux, L. Bruno, S. Gardiner, and J. C. Downs. "Loss of Elasticity in the Aging Human Sclera." *Invest. Ophthalmol. Vis. Sci.* 53, no. 14, 2800. http://iovs.arvojournals.org/article.aspx?articleid=2187713.

Q184 **How does aging affect the cornea?**

A184 The lacrimal glands produce less and less tears with age, causing dry corneas that, if untreated, could lead to corneal infection and inflammation. Furthermore, it has been documented that in individuals with astigmatism, the corneal curvature gradually changes from with-the-rule to against-the-rule.

Asano, Kazuko, Hideki Nomura, Makiko Iwano, Fujiko Ando, Naoakira Niino, Hiroshi Shimokata, and Yozo Miyake. "Relationship between astigmatism and aging in middle-aged and elderly japanese." *Jpn. J. Ophthalmol.* 49, no. 2 (2005): 127–33. Accessed September 10, 2017. doi:10.1007/s10384-004-0152-1.

Q185 **How does aging affect the iris and the pupil?**

A185 Aging weakens the iris muscles that control the size of the pupil. As a result, the pupil becomes smaller and less able to open in response to dim environments. Typically, someone about 60 years of age requires three times more light to see properly than people in their 20s.

"Your Aging Eyes." *National Institutes of Health.* September 07, 2017. Accessed September 10, 2017. https://newsinhealth.nih.gov/issue/Jan2011/Feature1.

Q186 **How does aging affect the lens?**

A186 One major effect of aging on the lens is cataract development, which can lead to loss in vision. However, the three major proteins of the lens, alpha-, beta-, and gamma-**crystallins**, have a constant risk throughout the aging process of other deleterious changes. For example, deamidation and glycation of crystallin proteins cause alterations in the structure of crystallin proteins, causing them to be broken down into protein fragments. The accumulation of these protein fragments causes the decrease in transparency of the lens, especially in the blue wavelengths, leading to an orange-brown tint, and in extreme cases, cataracts. This process also causes the hardening and decrease in flexibility of the lens, leading to presbyopia. Furthermore, the lens changes shape slower, leading to it taking longer to focus properly on objects of varying distances.

Michael, R., and A. J. Bron. "The ageing lens and cataract: a model of normal and pathological ageing." *Philos. Trans. R. Soc. Lond., B, Biol. Sci.* 366, no. 1568 (2011): 1278–292. Accessed September 10, 2017. doi:10.1098/rstb.2010.0300.

Sharma, K. Krishna, and Puttur Santhoshkumar. "Lens aging: Effects of crystallins." *Biochim. Biophys. Acta.* 1790, no. 10 (2009): 1095–108. doi:10.1016/j.bbagen.2009.05.008.

Q187   What are cataracts and why do they occur?

A187   A cataract is the clouding of the lens which adversely affects vision. The proteins of the eye are arranged in a way that keeps the lens clear, but when these proteins begin to clump together they cloud the lens, causing cataracts. Cataracts usually begin very small, so small that they are almost unnoticeable, and slowly progress to eventually significantly affect vision. Cataracts can severely reduce the sharpness of images and add a brownish tint to them. By age 75, half of all caucasians have cataracts. Usually beginning at the age of 50, the cataract is unnoticeable and becomes a detriment to their vision beginning in their 60s. To completely cure cataracts, a new artificial lens must replace the lens through surgery.

"Facts About Cataract." *National Eye Institute.* September 01, 2015. Accessed September 10, 2017. https://nei.nih.gov/health/cataract/cataract_facts.

Q188   Can cataracts return after surgery?

A188   After surgery cataracts cannot come back once it is removed, due to an artificial lens replacing the individual's natural lens. However, one may still experience a blurriness that mimics the symptoms of a cataract known as **posterior capsular opacification**, but this condition is easily treatable through a simple laser procedure. Posterior capsular opacification is simply when the lens capsule, which holds the artificial lens in place, thickens, making one's vision cloudy.

Ray, C. Claiborne. "Through a Glass, Darkly." *The New York Times.* April 01, 2013. Accessed September 10, 2017. http://www.nytimes.com/2013/04/02/science/can-cataracts-grow-back-after-surgery.html.

Q189 **What is presbyopia and how is it treated?**

A189 **Presbyopia** is a common eye condition that commonly begins to occur around the age of 40. This condition is characterized by difficulty focusing on close objects, and is caused by aging. If someone needs to hold a menu at an arm's length away, it is likely he/she has presbyopia. Presbyopia is thought to be caused by the loss of flexibility of the lens of the eye and gradual thickening as people grow older.

Treatment for presbyopia often simply requires the prescription of appropriate eyeglasses. Bifocal lenses are optimal for presbyopia, as they contain two lenses, on the bottom for near vision, and on the top for far vision. Another common option people use is reading glasses which have a single lens and are solely used for close activities such as reading. Yet another treatment is monovision. **Monovision** is when one eye uses lenses to perceive far objects far away, and one eye views close objects. This option however is not always recommended due to the decrease in the ability to perceive depth, as it is necessary to use both eyes for stereopsis.

Another way to treat presbyopia is through surgery. Once common method is through **refractive lens exchange** (RLE) which replaces the patient's lens with artificial **intraocular lens** (IOL) to improve vision. Both multifocal IOL and accommodating IOL help with clear vision in both distance and near vision without the need of using glasses.

Bailey, Gretchyn. "One Vision Problem You Can't Avoid." *All About Vision.* August 2017. Accessed September 10, 2017. http://www.allaboutvision.com/conditions/presbyopia.htm.

"Surgery for Presbyopia." *All About Vision.* July 2017. Accessed September 10, 2017. http://www.allaboutvision.com/visionsurgery/presbyopia_surgery.htm.

Q190 **What are bifocal and progressive lenses?**

A190 **Bifocal** and progressive lenses are two types of lenses used to treat presbyopia. In bifocals, the lens is divided into two sections with different refractive powers, one for distance vision and one for near vision, to compensate for the lack of accommodation in presbyopia. Similarly, trifocal lenses have three regions, for near, intermediate, and far vision.

However, a problem encountered by these bifocals and trifocals is image distortion when an object is viewed on the border between two regions due to different refractive powers. A progressive lens is not divided into distinct regions, but has a gradient that gradually transitions from low to high refractive power.

Heiting, Gary. "Bifocals & Trifocals Solutions." *All About Vision*. August 2017. Accessed September 10, 2017. http://www.allaboutvision.com/lenses/multifocal. htm.

Heiting, Gary, and Mark Mattison-Shupnick. "The Advantages of Progressive Lenses Over Bifocals and Trifocals." *All About Vision*. August 2017. Accessed September 10, 2017. http://www.allaboutvision.com/lenses/progressives.htm.

Q191 **How does aging affect the retina?**

A191 A variety of changes occur in the retina as a person gets older. The acuity and sensitivity of the peripheral visual field decreases and the retina becomes less sensitive under low-light, or scotopic conditions. Furthermore, the retina requires more time to adjust between different lighting conditions, such as walking from a brightly lit room to a dark room, or vice versa.

One of the main reasons behind these changes is the death of photoreceptor cells. Because these cells are neurons, when they die, they are not replaced, and so the number of photoreceptors gradually decreases over time. This seems to affect rods more than cones, as the number of rods in the eye decreases about 50% between 20 and 40 years of age, which is about 970 rods/mm²/year. The higher rate of rod death than cone death explains why aging affects scotopic more than photopic vision, as rods are crucial for low light vision.

A cause of the death of photoreceptor cells is the formation of chemicals called **free radicals**. As a person ages, chemicals collectively known as lipofuscin is deposited in the retina, which is almost continuously subjected to light, a form of energy. The constant input of energy forms a type of chemical stress causing the lipofuscin to form these free radicals, such as reactive oxygen species. Examples of reactive oxygen species include peroxides (O_2^{2-}), superoxides (O_2^-), and singlet oxygen. Free radicals are highly reactive and can trigger apoptosis, or cell death, in photoreceptors and other cells, as well as causing damage to DNA, RNA and proteins.

While many of the effects of aging, such as decreased light sensitivity, are gradual and do not severely impair vision, increased age also increases the risk for getting several serious eye conditions, such as age-related macular degeneration, retinal detachment, and glaucoma.

Bonnel, Sébastien, Saddek Mohand-Said, and José-Alain Sahel. "The aging of the retina." *Exp. Gerontol.* 38, no. 8 (2003): 825–31. doi:10.1016/s0531-5565(03) 00093-7.

"NCI Dictionary of Cancer Terms." *National Cancer Institute.* Accessed September 10, 2017. https://www.cancer.gov/publications/dictionaries/cancer-terms?cdrid=687227.

Q192 **What are the effects of aging on seeing in the dark?**

A192 Difficulty seeing under low-light, or scotopic, conditions is common among the elderly. Unlike under photopic conditions, the decrease in sensitivity under scotopic conditions is largely due to neural deterioration. When one ages, the density of rods and ganglion cells is decreased, which possibly contributes to the difficulty seeing in the dark in comparison to younger people. However, the mechanisms of the deterioration of the neurons in the visual pathway is still under debate. Another possible explanation for the decline in scotopic vision with age may be that rhodopsin regeneration slows down as one becomes older.

Owsley, C. "Aging and Vision." *Vision Res.* 51, no. 13 (2011): 1610–622. doi:10.1016/j. visres.2010.10.020.

Q193 **What is age-related macular degeneration?**

A193 **Age-related macular degeneration**, or AMD, is a disease where the macula of the retina degenerates with age, leading to the loss of vision at the center of the visual field. There are two types of age-related macular degeneration, dry and wet macular degeneration. Dry AMD, the more common type found in 90% of people with AMD, occurs when deposits of the fatty protein drusin begin to occur in the macula and gradually the macula becomes thinner and eventually stops working. This type of AMD is slowly progressive. With wet AMD, found in the remaining 10% of AMD patients, vision loss is usually more sudden and significant. This type is more dangerous and currently has no treatments. With wet AMD, **choroidal neovascularization** occurs, which is the growth abnormal blood vessels from the choroid under the retina. These new blood vessels can leak fluid, blurring vision. This form of AMD progresses much faster the dry AMD and is much more noticeable. Treatment for wet AMD includes anti-VEGF treatment, which uses medication to block vascular endothelial growth factors, and thermal laser treatment, which cauterizes ocular blood vessels.

"Facts About Age-Related Macular Degeneration." *National Eye Institute.* September 01, 2015. Accessed September 10, 2017. https://nei.nih.gov/health/maculardegen/armd_facts.

Q194 **What are floaters?**

A194 Floaters are small particles within the vitreous humor that can refract or scatter light as it passes through the eye, causing visual artifacts such as spots and lines. With increased age, the vitreous humor shrinks, causing the collagen that makes up the vitreous humor to break down into debris that gradually increases the amount of floaters. This process is normal and generally does not impair vision. However, a sudden increase in the amount of floaters could be a result of retinal detachment, a serious condition that should be checked immediately.

"Floaters." *Department of Ophthalmology.* Accessed September 10, 2017. http://eye.osu.edu/eye-conditions/comprehensive/floaters/index.cfm.

Q195 **What is retinal detachment?**

A195 When the movement of the vitreous fluid of the eye pulls harder than normal on the retina, the retina may tear, allowing for fluid to enter the hole and lift the retina off the back of the eye like wallpaper peeling off a wall. This is known as **retinal detachment**. A sudden increase in the amount of floaters and flashes, a shadow appearing in the periphery, and a sudden decrease in visual acuity are common symptoms of the initial stages of retinal detachment. An ophthalmologist can diagnose retinal detachment when he or she dilates the pupils. Immediate action should be taken to fix a retinal detachment since it will most likely result in permanent blindness. A retinal detachment can be fixed using laser surgery (photocoagulation), freezing treatment (**cryopexy**), or by scleral buckle surgery.

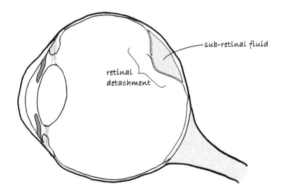

In retinal detachment, the retina is detached from the underlying retinal pigment epithelium. Retinal detachment can lead to permanent blindness.

Some common causes of retinal detachment include injury and diabetes. Furthermore, aging makes the vitreous humor of the eye more liquid. This causes the vitreous humor to separate from the surface of the retina. As the humor separates, it may force the retina to pull apart, causing retinal detachment.

"Facts About Retinal Detachment." *National Eye Institute.* October 01, 2009. Accessed September 10, 2017. https://nei.nih.gov/health/retinaldetach/retinaldetach.

Q196 **What is glaucoma?**

A196 **Glaucoma** is an eye condition that leads to damage to the optic nerve. Glaucoma can cause blindness within a few years if left untreated. If you are over 40 and have a family history of glaucoma, it is advised to get regular checkups from eye doctors.

Glaucoma is caused by an increase in **intraocular pressure**, or the pressure within the eye, due to the buildup of fluids in the eye.

In a normal eye, aqueous humor, a liquid in the front of the eye, is constantly being created by the ciliary body in the posterior chamber and being drained out through a vessel known as **Schlemm's canal**, which is covered by a structure called the **trabecular meshwork**. If the trabecular meshwork somehow gets clogged, there is be a buildup of aqueous humor, which thus increases the intraocular pressure. Alternatively, the increase in intraocular pressure could be a result of increased aqueous humor production. In either case, this pressure pushes back the lens, which then puts pressure on the vitreous humor of the eye and the blood vessels and nerves in the back of the cell, including the optic nerve. The reason of this blockage is often, but age and genetics appear to be risk factors for developing glaucoma. Other causes of glaucoma, although uncommon, include eye trauma, eye infection, inflammation, or blocked blood vessels. There are two main types of glaucoma, open-angle and angle-closure glaucoma.

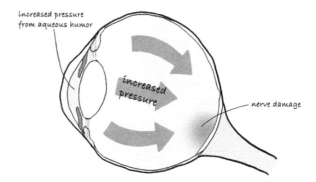

In glaucoma, a buildup of aqueous humor in the anterior of the eye leads to increased pressure on the vitreous humor, which in turn causes pressure on the optic nerve. This leads to nerve damage and potentially vision loss.

Open-angle glaucoma, also known as chronic glaucoma, is when the trabecular meshwork is clogged somehow. This results in gradual increase of intraocular pressure, leading first to impairment of peripheral vision, then central vision. Early detection of glaucoma is important, as once a part of vision is impaired it cannot be restored.

Angle-closure glaucoma, also called acute glaucoma, is a result of blockage of the trabecular meshwork because the angle between the iris and the cornea is too narrow. In acute angle-closure glaucoma, this process occurs rapidly and can create a dramatic increase of intraocular pressure leading to vision loss. This condition requires immediate medical attention. Fortunately, angle-closure glaucoma is less common than open-angle glaucoma.

"What Is Glaucoma?" *WebMD.* Accessed September 10, 2017. http://www.webmd.com/eye-health/glaucoma-eyes#1.

"Intraocular Pressure." *American Academy of Ophthalmology.* Accessed September 10, 2017. https://www.aao.org/bcscsnippetdetail.aspx?id=f010bbf6-3f3e-486b-b5cd-0ad86ddb9d74.

"Glaucoma Risk factors." *Mayo Clinic.* September 15, 2015. Accessed September 10, 2017. http://www.mayoclinic.org/diseases-conditions/glaucoma/basics/risk-factors/CON-20024042.

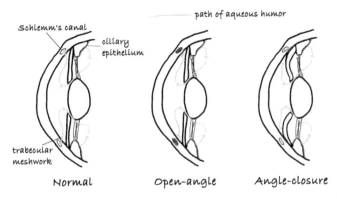

In a normal eye, aqueous humor is produced by the ciliary epithelium and drained through the trabecular meshwork into Schlemm's canal. In open-angle glaucoma, the trabecular meshwork is blocked, leading to a buildup of aqueous humor. In angle-closure glaucoma, the iris and cornea press together, preventing the aqueous humor from being drained.

Q197 **How is glaucoma treated?**

A197 Treatment options for glaucoma aim to alleviate pressure from the vitreous humor onto the optic nerve (intraocular pressure). The first choice of treating glaucoma is usually eye drops. These eye drops effectively reduce intraocular pressure, and thus reduces damage done to the optic nerve. There are multiple types of eye drops based on its main ingredient. These include eye drops containing chemicals known as prostaglandins, beta-blockers, alpha-adrenergic agonists, and carbonic anhydrase inhibitors. Certain people are poor candidates for eye drops, however, for it may worsen their condition. In this case, surgery might be recommended. For appropriate treatment, it is imperative to consult a doctor.

Haddrill, Marilyn. "Glaucoma Treatment: Eye Drops and Other Medications." *All About Vision*. April 2016. Accessed September 10, 2017. http://www.allaboutvision.com/conditions/glaucoma-3-treatment.htm.

Q198 How can diabetes cause damage to your vision?

A198 Chronic **hyperglycemia**, caused by diabetes, can gradually damage the tiny blood vessels in the retina, which is known as diabetic retinopathy. Diabetic retinopathy can cause the blood vessels to hemorrhage, distorting vision. In very late stages, diabetic retinopathy can cause the growth of abnormal blood vessels in the retina causing scarring and cell loss. Having diabetes increases the risk of getting diabetic retinopathy. Diabetic retinopathy can cause vision loss in two ways: macular edema or proliferative retinopathy.

Macular edema is caused by tiny leaks in the retinal blood vessels, causing blood and fluid to leak out of the vessels and deposit fatty material in the retina. This causes the macula to swell and blur the patient's vision. In order to treat this condition, Anti-VEGF and corticosteroid treatment are prescribed. If the case is more serious a variety of surgical procedures may be used. An ophthalmologist will use focal laser treatment to reduce swelling of the macula. The procedure stabilizes vision by sealing of leaking blood vessels that interfere with the proper functioning of the macula by applying many small laser pulses to regions of leaking fluid.

Proliferative retinopathy occurs when abnormal blood vessels begin to grow on the retina's surface. The new blood vessels are very fragile capillaries that bleed very easily leading to the formation of scar tissue. This scar tissue can then contract causing retinal detachment which can lead to blindness. A vitrectomy, in which some of the vitreous gel is surgically removed, can be used to treat severe bleeding. Scatter laser surgery, in which 1,000 to 2,000 very small laser burns are made, cause the abnormal blood vessels to shrink.

"Hyperglycemia in diabetes." *Mayo Clinic*. April 18, 2015. Accessed September 10, 2017. http://www.mayoclinic.org/diseases-conditions/hyperglycemia/basics/definition/con-20034795.

Q199 What is temporal arteritis?

A199 Temporal arteritis, also known as giant cell arteritis, is a disease of unknown cause characterized by the inflammation of arteries. One major risk factor of temporal arteritis is age, as the disease usually affects people older than 50 years of age. It typically causes inflammation of the arteries of the temporal region of the head, but can also affect other arteries of the head. This leads to various symptoms including headache, jaw pain, and fatigue. In approximately 20% of patients, temporal arteritis leads to some form of permanent visual damage and even blindness of both eyes in the most severe cases, resulting from the loss of blood flow to the eyes. Temporal arteritis cannot be cured, but is treated through corticosteroids, which reduce inflammation and restore blood flow to the eyes.

Diamond, H. S., and M. Seetharaman. "Giant Cell Arteritis (Temporal Arteritis)." *Medscape*. August 29, 2017. Accessed September 10, 2017. http://emedicine.medscape.com/article/332483-overview.

Q200 What can be done to reduce the risk of developing eye diseases with age?

A200 It is possible to protect your eyes and decrease the risk of developing age-related eye diseases. Light, while allowing us to see, can also be damaging to the eyes, causing the creation of harmful free radicals. Ultraviolet light is especially damaging, as these light rays carry high amounts of energy. Sunglasses are recommended to block out ultraviolet rays on sunny days.

Furthermore, antioxidants are a class of chemicals, including vitamins A, C, and E, that oppose free radicals and neutralize them, preventing them from damaging cells. A study known as the **AREDS** (Age-Related Eye Disease Studies) conducted by the National Eye Institute found that taking antioxidants could help prevent eye disease. In a group of people at high risk for age-related macular degeneration, antioxidant supplements decreased vision loss by 19% and helped slow the progression of the disease. Other studies have also found that taking antioxidants like vitamin C can help reduce the risk for cataracts. The NIH Office of Dietary Supplements recommends that people over the age of 30 take at least 700 micrograms of vitamin A, 75 milligrams of vitamin C, and 15 milligrams of vitamin E daily.

"Antioxidants & Age-Related Eye Disease." *American Optometric Association*. Accessed September 10, 2017. https://www.aoa.org/patients-and-public/ caring-for-your-vision/nutrition/antioxidants?sso=y.

"What is AREDS?" *National Eye Institute*. October 2004. Accessed September 10, 2017. https://web.emmes.com/study/areds/.

Ross, A. C, C. L. Taylor, and A. L. Yaktine. "Dietary Reference Intakes for Calcium and Vitamin D." *Institute of Medicine*. January 01, 1970. Accessed September 10, 2017. https://www.ncbi.nlm.nih.gov/books/NBK56068/.

Afterword

Q201 **Do you see with your eyes or with your brain?**

A201 The visual system is an extremely complicated system, encompassing a variety of structures ranging from the cornea to the various cortical visual areas of the occipital, parietal, temporal, and frontal lobes. Light entering the eye is just the beginning of the process of seeing.

The cornea and lens together form a focused image on the retina, which contains the photoreceptors that detect light. The photoreceptors, rods and cones, of the retina detect photons and relay the information about them (including location and wavelength) to bipolar and ganglion cells, with various other interneurons serving to enhance the quality of the visual signal.

While the eye is clearly critical to the sense of vision, its function boils down to merely the detection of light; vision consists of much more complexity, a fact that is reflected in the large area of the brain dedicated to visual processing. The eye does not simply function as a sensor that creates the perception of seeing you and relays this perception to the brain. Instead, it encodes vision through edge information, from which the visual field is reconstructed by the brain, leading to the expression that the brain is responsible for seeing, not the eye. This can be demonstrated by various illusions, such as the Craik-O'Brien-Cornsweet illusion, which uses edge information to trick the brain into incorrectly reconstructing the image.

Through the neurons of the lateral geniculate nucleus, primary visual cortex, and subsequent extrastriate areas, visual information containing edges, colors, and orientations is processed and used to create the perception of seeing. The opponent process calculates the hue of an object, while the orientation-sensitive cells of the cortex put edges together to form identifiable objects. Illusions take advantage of this reconstruction process to alter the perception of reality. Afterimages that do not actually exist are seen as a result of neuronal fatigue of cells in the brain. The object seen in the Rubin vase illusion is sometimes a vase and sometimes two faces as a result of ambiguous figure-ground segregation. On the other hand, color constancy and double opponent cells demonstrate that the brain actively alters the perception of the world around you to maintain consistency.

The process of constructing a mental image of the world from light information is immensely complex. Sometimes, errors like illusions are harmless and fade away after minutes. Other times, however, issues of the visual system can persist and worsen as a result of many eye conditions. Due to the complexity of the visual pathway, abnormalities can arise from essentially any part of the visual system.

Damage to the cornea can lead to many major conditions that can adversely affect one's vision. Myopia, a disease that affects nearly one-fourth of the human population, is characterized by difficulty viewing distant objects clearly, but normal vision while viewing close objects. It is usually caused by an abnormal curvature of the cornea. Fortunately, this condition can be easily managed with glasses and contact lenses, or even corrected completely with laser eye surgery. Another condition, hyperopia, is similar to myopia, except the light is focused behind the retina instead of in front of it. This results in nearby objects looking blurry while maintaining normal acuity with far objects. Like myopia, hyperopia is also generally caused by errors in the curvature of the cornea.

The lens, like the retina, can also malfunction, resulting in a common condition known as cataracts. Cataracts occur where the lens of the eye becomes cloudy, causing blurred vision. Often, cataracts in children are caused by genetics or simply by eye trauma. To treat cataracts, many hospitals recommend surgery, as cataracts may gradually develop blindness or amblyopia (lazy eye).

After refraction by the cornea and the lens, light hits the photoreceptors of the retina, which converts it into electrical signals. However, if anything goes wrong with the photoreceptors, conditions like color blindness may occur. Color blindness occurs when one or more types of cone cells are either absent, nonfunctioning, or detect slightly different colors than they are suppose to.

Yet another structure of the visual pathway that could lead to a disease through malfunction is the trabecular meshwork. When the trabecular meshwork gets clogged, it can lead to glaucoma, an eye condition that leads to optic nerve damage due to the buildup of intraocular pressure. Glaucoma can cause blindness within a few years if left untreated.

Finally, disorders can also result from the tracts in the brain. When the optic radiations have a lesion, a disease known as quadrantanopia may occur. Quadrantanopia is when a certain quadrant of an individual's vision is foggy or indistinct relative to the other quadrants.

The visual system consists of an immense number of parts all working together to create this percept of seeing that we use every single day, beginning the moment we are born to the moment we die. From our journey through the visual system, we have examined the complexities of this amazing sense, and we hope that you continue to appreciate and take good care of your vision.

Glossary

A

accommodation—the process in which the lens changes curvature to allow the eye to focus on objects of varying depth. **Q8**

action potential—an impulse that travels along the cell's axon, transmitting the signal from one neuron to the next. **Q52**

age-related macular degeneration—a disease where the macula of the retina degenerates with age, leading to the loss of vision at the center of the visual field. **Q193**

amacrine cell—a type of retinal neuron located in the inner nuclear layer. There are many types of amacrine cells, and their functions are still being discovered. **Q61**

amblyopia—an eye condition in which the brain is unable to combine images from the two eyes. Amblyopia can be caused by various reasons, including strabismus, anisometropia, and cataracts. **Q178**

angle-closure glaucoma—a type of glaucoma resulting from the blockage of the trabecular meshwork because the angle between the iris and the cornea is too narrow. **Q196**

anisometropia—a condition where the two eyes have significantly different prescriptions. **Q178**

anterior chamber—the region of the eye between the cornea and the iris. It is filled with aqueous humor. **Q9**

Anton-Babinski syndrome—a condition that arises from occipital lobe damage where people are cortically blind but still believe that they are not blind and confabulate to fill in gaps caused by their lack of visual input. It is also known as visual anosognosia. **Q95**

aqueous humor—the fluid that fills the anterior and posterior chambers, providing support for the eye as well as providing nutrients for structures that have no blood vessels, like the cornea and lens. **Q9**

AREDS—a study conducted by the National Eye Institute investigating the age-related risk factors for macular degeneration and cataracts. **Q200**

astigmatism—a condition resulting from refractive error of the eye in which light is focused onto multiple points rather than a single point. **Q170**

autorefractor—an automated machine that is used to determine refractive error. **Q171**

axon—a long extension of the neuron that allows it to send signals across long distances, usually forming a synapse with another neuron. **Q52**

B

bacterial conjunctivitis—an inflammation of the conjunctiva caused by bacteria. **Q164**

Baum's loop—the optic radiation from the superior retina travelling through the retrolenticular limb of the internal capsule to the primary visual cortex. Damage to this loop can cause inferior quadrantanopia. **Q86**

bifocal lens—a type of glasses containing two regions with different refractive powers. **Q190**

binocular vision—the type of vision requiring both eyes. **Q159**

bipolar cell—a cell found in the retina that transmits information from rods and cones to ganglion cells. **Q57**

bistratified cells—a type of ganglion cell that has a large receptive field and projects to the koniocellular layers of the LGN. **Q73**

blindsight—a condition where visual reflexes are not lost, even though vision is lost. Occurs when visual damage only affects the striate cortex. **Q94**

blobs—the parts of the V1 which are sensitive to color and arrange in a cylinder pattern. Can be seen by a cytochrome oxidase stain. **Q108**

border ownership—the process of determining which edges belong to which object. Border ownership is exhibited in V2 and V4 as well as other areas. **Q129**

Brodmann's areas—a group of areas in the brain that are distinguished by their histology. There are 52 different areas on the cortex. **Q99**

C

cAMP—a chemical that stands for cyclic adenosine monophosphate and is used for signal transduction. **Q33**

carotenoid—a class of pigment that is present in the macula and other parts of the retina. These pigments include lutein and zeaxanthin which helps give the macula a yellowish color. **Q16**

cataract—the clumping of proteins in the eye clouding of the lens which adversely affects vision. **Q177**

center-surround antagonism—the opposing interaction between the center and surrounding regions of a receptive field of certain neurons in the visual pathway, such as retinal ganglion cells, allowing detection of edges. **Q56**

cerebral cortex—(or "cortex") a layer of neurons two to four millimeters thick on the surface of the brain composed of closely packed neurons. **Q97**

choroid—a vascular layer of tissue located between the sclera and retina. Provides all layers with approximately 90% of their blood supply. **Q19**

choroidal neovascularization—the abnormal growth blood vessels from the choroid under the retina. **Q193**

chromatic aberration—an optical property in which light rays of different wavelengths are refracted in different amounts. **Q47**

chromotopic map—a possible arrangement of globs in the cortex, where the position of the glob in the cortex corresponds to the hue to which it is most sensitive. **Q134**

ciliary body—a structure behind the iris consisting of the ciliary muscle and the ciliary epithelium. **Q7**

ciliary epithelium—a part of the ciliary body that produces aqueous humor to fill the anterior of the eye. **Q7**

ciliary muscle—a part of the ciliary body involved in accommodation and changing the curvature of the lens. **Q8**

circadian rhythm—an approximately 24-hour cycle pertaining to the physiological processes of an organism. External factors may contribute to the circadian rhythm. **Q93**

coloboma—a disease characterized by gaps in certain structures in the eye, such as the iris or the retina. **Q150**

color blindness—a condition characterized by a decreased ability to distinguish colors, occurring when one or more types of cone cells are either absent, nonfunctioning, or detect different colors than they are suppose to. **Q48**

color constancy—the ability of humans to perceive colors of objects regardless of the color of the light source. **Q115**

conjunctivitis—("or pink eye") a condition caused by certain pathogens, allergies, foreign objects, or chemicals in the eye. Common symptoms of the eye include redness, swelling, burning, tearing, crust-formation, and light-sensitivity. **Q164**

contact lens—a thin lens that is made of plastic. These lens are placed directly onto the eye usually to correct visual defects. **Q168**

contrast sensitivity function—a graph showing the threshold contrast level required to be visible as a function of the spatial frequency of the image. **Q181**

complex cells—a type of cell located in V1 that recognizes bars of lights that are in motion. **Q105**

cones—a cone-shaped photoreceptor that can detect color. There are three types of cones: red, blue, and green. **Q26**

cornea—the outermost layer of the eye which acts as a barrier against dirt, germs, and other harmful foreign particles. Furthermore, it plays a key role in focusing light onto the retina so that the image appears clear and sharp. **Q2**

corneal topographer—an instrument that measures the curve of the cornea, determining the smoothness of the surface and providing important information for degree of refractive error and diagnosing keratoconus. **Q171**

cover test—a common eye-test which helps diagnose strabismus (crossed eye) and amblyopia (lazy eye). During this test, the doctor covers one of the patient's eyes and instructs the patient to use the other eye to focus on a certain nearby object. **Q172**

Craik-O'Brien Cornsweet illusion—an illusion where the right-most edge of the image appears darker than the leftmost edge of the image. However, both edges are the same color. **Q64**

cranial nerve—a group of nerves that originate from the brain instead of the spinal cord. Humans have 12 cranial nerves. **Q179**

cryopexy—a type of treatment used for retinal detachment that uses intense cold to destroy retinal or choroidal tissue. **Q195**

crystallin—a transparent protein located in the cornea and lens. Breakdown of crystallins with age can lead to poor vision. **Q186**

cyclopia—a rare congenital disorder caused by a mutation in the sonic hedgehog gene that results in the formation of only one eye. **Q148**

CYL—the refractive power specified on eyeglass prescriptions to cure astigmatism. **Q174**

cytochrome oxidase—an enzyme part of the electron transport chain that is commonly used for staining of the visual cortex. **Q108**

D

dark adaptation—a process that allows the eye to become more sensitive to light under dim conditions. **Q37**

dendrite—a branching extension of a neuron that allows it to receive signals from other neurons. **Q52**

depth perception—the ability to use binocular vision, also known as stereopsis, along with other visual cues, to determine the distance to an object. **Q158**

deuteranomaly—a type of color-blindness in which the blue photopsin molecule malfunctions, diminishing blue-green color discrimination. **Q49**

diopters—a unit that measures the light-bending capabilities of a lens. **Q6**

dorsal stream—a pathway that runs from the primary visual cortex in the occipital lobe to various visual regions in the parietal lobe. This pathway is responsible for tracking and guiding movement by creating a detailed visual field, detecting and analyzing movement. **Q138**

double-opponent cells—a particular group of opponent cells whose centers are excited by one color and inhibited by the opposite of that color. **Q114**

E

early onset myopia—a type of near-sightedness that occurs in children, usually ages 9–11. **Q166**

ectoderm—the outermost primary germ layer found in the embryo. Later differentiates to form the epidermis, tooth enamel, and nervous system. **Q141**

ectropion—a condition where the lower eyelid to turn outward, exposing the inner eyelid. **Q182**

endoderm—the innermost primary germ layer found in the embryo. It later differentiates to form the lining of various organs in the body. **Q141**

end-inhibition—a property of hypercomplex neurons in which the response of the cell decreases as the size of the stimulus increases. **Q106**

entropion—a condition where the lower eyelid turns inward, causing the eyelid to rub against the eyeball. **Q182**

F

Fantz preferential looking method—an experimental technique used to infant vision, developed by Robert Fantz in 1961. **Q154**

fovea centralis—an area in the retina that contains closely-packed cones. It is located in the center of the macula and allows for sharp central vision. **Q13**

free radical—a type of highly reactive chemical that can trigger apoptosis and cause damage to DNA, RNA and proteins in photoreceptors and other cells. **Q191**

frontal lobe—the lobe of the cerebrum responsible for emotions, personality, problem solving, motor skills, judgement, and social behaviour. **Q98**

G

GABA—the most common inhibitory neurotransmitter. **Q151**

ganglion cell—a cell found in the retina which has axons that make up the optic nerve. **Q62**

ganglion cell layer—the inner layer of the retina that contains ganglion cells. **Q23**

glaucoma—an eye condition that causes damage to the optic nerve due to excess aqueous humor in the anterior chamber. **Q196**

globs—a group of cells in the V4 that are color sensitive. They are similar to the stripes in V1 and V2. **Q134**

glutamate—the most common excitatory neurotransmitter in the nervous system. **Q34**

Goldmann tonometer—a device used to measure intraocular pressure (IOP) and helps check for glaucoma. **Q171**

G-protein coupled receptor—a type of receptor that is involved in signal transduction. **Q33**

great-grandmother cell hypothesis—an early theory that was created to try and explain visual processing. It is now known to be false. **Q96**

H

habituation method—a method of experimenting with visual perception in infants. An infant is presented with a first stimulus for a certain period of time. After the infant becomes use to the stimulus, known as habituation, it is replaced with the second stimulus. If the infant immediately spends more time looking at the second stimulus, then the visual system is capable of distinguishing the two stimuli. **Q155**

Helmholtz Hypothesis—a hypothesis that describes the mechanism of accommodation. It states that the ciliary muscle pull on the zonular fibers when relaxed. This tension on the zonular fibers pulls the lens into a flatter form. And when the ciliary muscle contracts, tension on the zonular fibers decreases, allowing the lens to become more round. **Q8**

higher-order visual areas—the extrastriate cortical areas where processing of color, shape, and motion takes place. These areas include: V2, V3, V4, and V5. **Q121**

homonymous hemianopsia—a type of lesion in the optic tract that leads to visual field loss on either the right or left of the vertical midline. **Q79**

horizontal cells—a type of neuron found in the retina responsible for lateral interactions and formation of receptive fields. They provide negative feedback to the photoreceptors allowing the eye to see well in bright and dim light through adapting to the lighting condition by inhibiting photoreceptors. **Q60**

hypercomplex cell—(or "end-stopped cell") a group of cells that demonstrate a property known as end-stopping, in which the response of the cell decreases as the size of the stimulus increases. **Q106**

hyperglycemia—an excess of sugar in the blood that may be caused by diabetes. This can damage blood vessels of the retina. **Q198**

hyperopia—an eye condition common in children also called far-sightedness when the light is focuses behind the retina instead of in front of it. This results in blurry images for near objects while maintaining normal acuity with far objects. **Q168**

I

"ice-cube" model—a model that represents how the ocular dominance columns and orientation columns run perpendicularly, creating a grid in the primary visual cortex. **Q120**

illusory contours—a type of visual illusion where contours, or edges, are perceived where there are none, due to the positioning and arrangement of other shapes in an image. **Q128**

inner nuclear layer—the middle layer of the retina that contains amacrine, horizontal, and bipolar cells. **Q23**

inner plexiform layer—the layer of the retina between the inner nuclear and ganglion cell layers that contains synapses between neurons of the two layers. **Q23**

inner segment—the segment of the photoreceptor cell that contains the ribosomes and mitochondria. **Q42**

interblob—the regions in the primary visual cortex that do not stain from cytochrome oxidase. **Q116**

interglobs—the regions in V4 between globs. **Q135**

inter-stripes—the parts of the V2 that are thought to function for the perception of form. **Q127**

intraocular lens—an artificial lens that replaces a patient's own lens to improve vision in patients with presbyopia. **Q189**

intraocular pressure—the fluid pressure inside the eye. **Q196**

intrinsically photosensitive ganglion cell—a type of neuron that is the only retinal cell, other than rods and cones, that is photosensitive. **Q74**

iris—a circular muscle that surrounds the pupil that can be dilated and constricted to regulate the amount of light entering the eyes by using the sphincter muscles to constrict the pupil and dilator muscles to dilate it. **Q3**

K

Kanizsa triangle—a type of visual illusion that creates the perception of a triangle formed from illusory contours. **Q128**

keratometer—a machine used to measure the curve of the cornea, providing important information about the degree of refractive error. **Q171**

koniocellular cell—a type of neuron in the lateral geniculate nucleus whose specific function is unknown. **Q85**

L

L cones—the type of cones that detect red light. The L corresponds to the long wavelength of red light. **Q39**

LASIK—a procedure in which a corneal flap is created to correct myopia (nearsightedness). **Q175**

lateral geniculate nucleus—a nucleus found in the thalamus that is the midway station for most visual processing. Contains magnocellular, parvocellular and koniocellular cells which receive rods, red and green cones, and blue cones respectively. **Q81**

lateral intraparietal sulcus—a sulcus containing the lateral intraparietal cortex which is involved in saccades. **Q138**

lens—the part of the eye that focuses light. It is located between the posterior and anterior chamber. **Q6**

lutein—a pigment that is found mainly in the macula, it functions to protect the eyes from oxidative stress and high energy photons. **Q16**

M

M cones—the type of cones that detect green light. The M stands for medium, indicating the certain wavelength they are sensitive to. **Q40**

Mach bands—a type of illusion named after Ernst Mach. In the image, each rectangle appears lighter on the left side than on the right side, even though each is of uniform color. **Q63**

macula lutea—a yellowish area located near the center of the retina that contains the fovea. **Q15**

macular edema—a way diabetic retinopathy can lead to loss of vision. Macular edema is caused by tiny leaks in the retinal blood vessels, causing blood and fluid to leak out of the vessels and deposit fatty material in the retina. This causes the macula to swell and blur the patient's vision. **Q198**

magnocellular cells—a group of cells in the lateral geniculate nucleus. They receive input from parasol ganglion cells of the retina, and thus are part of the parasol system. **Q84**

melanin—the pigment that is present in the iris pigment epithelium and iris stroma responsible for eye color. **Q4**

Melanopsin—the photopigment present in intrinsically photosensitive retinal ganglion cells (ipRGC), allowing the cells to distinguish light. **Q93**

membrane potential—the difference of electric potential between the inside of the cell and the surrounding extracellular matrix, important for transmitting information through neurons. **Q52**

membranous discs—the part of a photoreceptor cell located in the outer segment that gives these cells their unique rod and cone shape. **Q26**

Meyer's loops—the two loops that carry information from the lower half of the retina and form the lower division of the optic radiation. **Q86**

mesoderm—the layer of the human embryo that develops into muscles, bones, and blood cells. **Q141**

microsaccade—the little jumps eyes make when viewing an object. **Q106**

midget cells—a type of retinal ganglion cell that project to the parvocellular layers of the lateral geniculate nucleus. They are called midget cells due to their small size in cell body and dendrites. **Q71**

midget system—a pathway stemming from the midget ganglion cells of the retina that extends to the various areas of the cerebral cortex. **Q75**

monochromacy—rare type of color blindness where the person with this condition can only see in black and white. **Q49**

monovision—a treatment option for presbyopia, when one eye uses lenses to perceive far objects far away, and one eye views close objects. However, this treatment may reduce stereopsis. **Q189**

motion aftereffect illusion—an illusion in where the viewer fixates at a spot on the screen while there is constant motion in a single direction. After a period of time, the user looks away from the screen and sees distortions near the center of the field of vision, which gradually fades away over a period of about a minute. **Q107**

multifocal lens—a type of lens that contains multiple prescription on a single lens, each prescription used for a certain distance. **Q169**

medial temporal—(or "MT") a region in the brain that receives input from V1, V2, and V3 as well as the koniocellular regions of the LGN. The MT is important in determining the speed and direction of a moving object, as well as movement of the eye. **Q136**

Myopia—a common condition also known as nearsightedness. Myopia is characterized by difficulty viewing distant objects clearly, but normal vision while viewing close objects. **Q166**

N

neurotransmitter—a substance that is released by synaptic terminals to transmit information from one neuron to another. **Q52**

O

ocular dominance column—a group of columns of neurons in the striate cortex that respond to input from one eye preferentially to the other. **Q118**

open-angle glaucoma—a type of glaucoma when the trabecular meshwork is clogged. This results in gradual increase of intraocular pressure, leading first to impairment of vision. **Q196**

opsin—light-sensitive proteins found in the photoreceptors of the retina. **Q33**

optic chiasm—an area where the two optic nerves cross over below the brain. **Q77**

optic disc—the raised area of the retina where the optic nerve enters the eye, lacking photoreceptors and thus resulting in a blind spot. **Q11**

optic radiation—the group of axons spanning from the LGN to the striate cortex. **Q86**

optic tract—the continuation of the optic nerve that spans from the optic chiasm to the LGN, pretectal nuclei, and the superior colliculus. **Q78**

optic nerve—the second pair of cranial nerves that transmit impulses from the retina to the brain. **Q11**

opponent theory—a theory that suggests that color perception works upon three opposing color systems: black and white, yellow and blue, and red and green. **Q109**

orientation column—the regions of the striate cortex that contain neurons that respond to visual input of varying angles. **Q119**

outer nuclear—the outer layer of the retina that contains the soma of photoreceptors. **Q23**

outer plexiform layer—the layer of the retina between the outer and inner nuclear layers that contains synapses between neurons of the two layers. **Q23**

outer segment—the modified cilia in photoreceptor cells that are responsible for absorbing light. **Q31**

ON bipolar cell—a type of bipolar cell that functions during the day-time. Found in the inner layer of the inner plexiform layer of the retina. **Q66**

ON channel—the congregation of bipolar and retinal ganglion cells that help detect contrast when the visual field is brighter than the surrounding. **Q65**

OFF bipolar cell—a type of bipolar cell that functions during the night. Found in the outer layer of the inner plexiform layer of the retina. **Q67**

OFF channel—the congregation of bipolar and retinal ganglion cells that help detect contrast when the visual field is darker than the surrounding. **Q65**

P

parietal lobe—the lobe of the cerebrum that processes information related to several senses, including taste, temperature, touch, and some vision. **Q98**

PAX6—a gene encoding a transcription factor involved in the development of the eye. It can be described as the "master" gene for eye development. **Q145**

persistent pupillary membrane—a condition where the fetal membrane persists in tissue spanning across the pupil, making the pupil appear segmented. **Q149**

photobleaching—the reduction in sensitivity of a photoreceptor following exposure to light, due to the conversion of 11-cis-retinal into all-trans-retinal. **Q37**

phoropter—a device with multiple lenses through which an eye chart can be viewed. It is used to determine the refractive error and prescription. **Q171**

photopsin—a group of photopigments, similar to rhodopsin, that are responsible for absorbing and detecting light in cones. Each type of cone (red, green, and blue) have a distinct type of photopsin which is most sensitive to a certain wavelength of light. **Q39**

photoreceptor—a type of neuron that is located in the retina. These cells convert light into chemical and electrical signals that are passed on to other neurons. **Q25**

photorefractive keratectomy—a type of laser eye surgery used to correct refractive errors in the cornea. PRK differs from LASIK in that the corneal epithelium is removed rather than creating a corneal flap. **Q175**

posterior capsular opacification—a condition where the lens capsule, which holds the artificial lens in place following refractive lens exchange, thickens, making one's vision cloudy. **Q188**

posterior chamber—the region between the iris and the zonule of Zinn. It is filled with aqueous humor. **Q9**

presbyopia—a common eye condition caused by aging that is characterized by difficulty focusing on close objects. **Q189**

pretectal area—a region of the brain that receives information from the retinal ganglion cells via the optic tract and is involved in the pupillary light reflex. **Q90**

primary visual cortex—the part of the occipital lobe that is the location of the first stage in processing visual information. **Q100**

protanomaly—a type of color-blindness where the absorption spectrum of the red photopsin is shifted toward the green photopsin, diminishing the ability to distinguish between red and green colors. **Q49**

Purkinje effect—the reduction of the perception of red colors under dim conditions. It is named after Jan Evangelista Purkinje and is caused by the lower sensitivity of rods to red wavelengths. **Q36**

pupil—a hole at the center of the iris through which light enters the eye. Its size depends on the contraction muscles in the iris. **Q5**

pupillary light reflex—a reflex where the size of the pupil changes depending on the amount of light available. **Q91**

Q

quadrantanopia—the loss of vision in a specific quadrant of vision. Occurs because of lesions in optic radiation. **Q87**

R

Rayleigh scattering—the scattering of light that is responsible for blue eye color as well as the color of the sky. **Q4**

receptive field—the area of the visual field in which the cell responds to stimulus. **Q53**

refractive lens exchange—a treatment for presbyopia which replaces the patient's lens with artificial intraocular lens. **Q189**

refractive power—(or "optical power") the quantification the curvature of the cornea in diopters. This measures the degree to which any optical system converges or diverges light. **Q165**

retina—the part of the eye composed of neural tissue that covers around two-thirds of the back of the eye. It is responsible for the detection of light. **Q12**

retinal—a chemical derived from vitamin A. This chemical is attached to a protein to form rhodopsin. **Q32**

retinal detachment—a condition when the movement of the vitreous fluid of the eye pulls harder than normal on the retina, the retina may tear, allowing for fluid to enter the hole and lift the retina off the back of the eye like wallpaper peeling off a wall. **Q195**

retinal disparity—the way one's left and right eye view slightly different images. The two different images "blend" to make you see one image, and also helps with depth perception. **Q126**

retinal ganglion cells—a group of cells in the retina that are located in the ganglion cell layer and receive input from photoreceptor cells through intermediate cells. **Q70**

retinal pigment epithelium—a layer of cells just outside the retina. The function of the RPE is to nourish the retinal cells and to participate in the regeneration of the photopigments in the retina. **Q18**

retinoscopy—an eye test used to determine which prescription of eyeglasses would be optimal for the patient. This test is performed in a dim room while the patient focuses on a large object, such as the large E on an eye chart. The doctor shines a light on the patient's eye and uses a machine to constantly switch the lens in front of the eye. The lens that is optimal for the patient's vision, helps determine the prescription needed for the patient. **Q172**

retinotopic map—a way of comparing how the parts of the retina that a certain image hits changes the areas of the brain that are stimulated when processing the image. **Q102**

rhodopsin—the photopigment present in rods that consist of protein attached to retinal. This helps the rods detect light. **Q32**

rods—a type of photoreceptor that are able to detect the brightness of light. The membranous discs of these photoreceptors are shaped to look like a rod, giving its name. **Q26**

Rubin vase—an optical illusion created by an ambiguous figure-ground segregation, leading to two possible foregrounds: the green area and the blue area. Depending on which is perceived to be the foreground, the viewer can either see a vase or two faces. **Q130**

S

Schlemm's canal—the vessel that drains out aqueous fluid from the eye. **Q196**

sclera—the surrounding white wall of the eyeball that maintains intraocular pressure. **Q20**

sensory transduction—the process in which the energy from the surrounding environment is converted to electrical impulses by receptors. **Q21**

simple cells—a group of neurons of the primary visual cortex with receptive fields similar to the center-surround receptive fields in the retina and LGN. **Q104**

slit lamp—an instrument that uses a high-intensity light source that shines a thin sheet of light into the eye to check the anterior portion of the eye for abnormalities including the conjunctiva, lens, iris, and cornea. **Q171**

Snellen chart—a chart containing letters with multiple rows of letters decreasing in size to test visual acuity. **Q173**

stereopsis—the perception of visual information from both eyes that used to determine the distances to object. **Q159**

strabismus—a condition in which the eyes are aligned incorrectly which is often caused by poor neuromuscular control of the eyes by weak or palsied cranial nerves responsible for eye movement. **Q179**

suprachiasmatic nucleus—a part of the hypothalamus that controls circadian rhythm. **Q92**

superior colliculus—a structure involved in visual reflexes that is located in the tectum of the midbrain. **Q89**

synapse—a junction where nerve cells connect to transmit information, usually through the use of neurotransmitters. **Q52**

T

tapetum lucidum—layer of tissue behind the retina that reflects back any light not initially absorbed by photoreceptors. **Q17**

temporal lobe—one of the four lobes of the brain, associated with visual memory and language recognition and comprehension. **Q98**

thalamus—mass of gray matter in the diencephalon of the brain that relays sensory and motor signals to the cerebral cortex. **Q81**

transcription factor—protein that regulates the rate of transcription. **Q144**

transducin—protein found in photoreceptors that is activated by the conformational change in rhodopsin when it interacts with light. **Introduction**

two-streams hypothesis—hypothesis that visual processing occurs in two distinct pathways: the ventral stream (what pathway) and dorsal stream (where pathway). **Q137**

V

V1—a part of the brain referred to as the primary visual or striate cortex. It contains a map representation of the visual field. **Q100**

V2—a part of the brain referred to as the secondary visual or prestriate cortex. It is the first region in the visual association area and has important functions in border ownership and figure-ground segregation. **Q122**

V3—a part of the brain directly in front of V2. It is primarily involved in orientation selection, motion and depth. **Q131**

V4—a part of the brain that is color sensitive and important in spatial vision. V4 makes up part of the extrastriate visual cortex along with V3 and V5. **Q132**

V5—a part of the brain that is anatomically defined as the middle temporal visual area. It is used mainly for motion and it can be divided into columns each tracking different directions of motion. **Q136**

Ventral Intraparietal Sulcus—one of 5 main regions of the intraparietal sulcus. The ventral intraparietal sulcus contains the ventral intraparietal area (VIP) which is involved in saccades. **Q138**

ventral stream—one of two streams in the Two Stream Hypothesis. It is also known as the "what" pathway and connects to the medial temporal lobe and the limbic system. Most of its input comes from the parvocellular layer of the LGN. **Q139**

visual acuity—the sharpness of the image that is perceived. **Q2**

visual field—the field of view of a person. **Q53**

vitamin A—a fat soluble vitamin that is converted into the chromophore for all visual opsin proteins. **Q41**

vitreous humor—a part of the eye that is found in the posterior chamber, it is a clear substance with a jelly-like consistency. **Q10**

Y

Young-Helmholtz Trichromatic Theory—a theory that proposed that there are three types of photoreceptors in the eyes. It further states that the three photoreceptors are blue cones (short-preferring), green cones (middle-preferring), and red cones (long-preferring), based on the wavelength of light that they responded to, and that these signals are interpreted as color by the brain. **Q46**

Z

zeaxanthin—a common carotenoid pigment that is naturally synthesized in plants and some other microorganisms. **Q16**

Zonule of Zinn—a ring of fibers that connects the ciliary body of the eye to the lens. Also known as suspensory ligaments. **Q6**

Index
With Question Numbers

CPSIA information can be obtained
at www.ICGtesting.com
Printed in the USA
LVHW011552170419
614526LV00027B/1644/P